The Short Guide Series

UNDER THE EDITORSHIP OF

Sylvan Barnet
Marcia Stubbs

A Short Guide
to Writing
about Biology

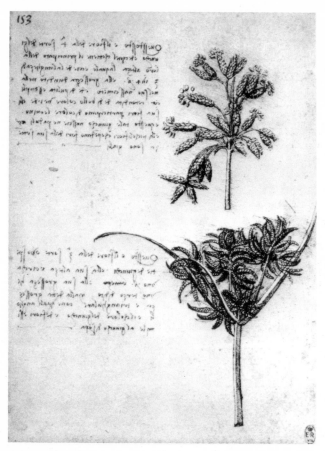

Leonardo da Vinci, Study of Rushes. Heads of two different types of rush. From the Royal Library at Windsor Castle. Copyright reserved. Reproduced by the gracious permission of Her Majesty Queen Elizabeth II.

A Short Guide to Writing about Biology

JAN A. PECHENIK
Tufts University

HarperCollins*Publishers*

Library of Congress Cataloging-in-Publication Data

Pechenik, Jan A.
  A short guide to writing about biology.

  Includes index.
  1. Biology—Authorship. 2. Report writing. I. Title.
QH304.P43 1987   808'.066574   86-10659

ISBN 0-673-39232-5

78910-MVN-969594939291

Acknowledgments
Figure 16 (p. 86) is reprinted with permission from the *Science Citation Index®*
1983 Annual. Copyright 1984 by the Institute for Scientific Information®,
Philadelphia, PA, U.S.A.

Figures 17, 18, and 19 (pp. 88, 89, and 90; from *Zoological Record*) and Figures
20 and 21 (pp. 91 and 93; from *Biological Abstracts*) are reproduced with the
permission of the copyright holder, B.A. Inc., 1986.

Figure 22 (p. 96) is reprinted by permission of the author and publisher. From
The Biological Bulletin 167 (August 1984): 200–202.

Pages 114–115: Proposal based on a paper by Acha Lord. Used by permission
of Charlotte (Acha) M. Lord.

Pages 173, 174, and 175: Big Mac® is a trademark of McDonald's Corporation.
Whopper® is a trademark of Burger King Corporation.

to Lindy
my wife, friend, editor, and fellow biologist

Preface

Biology instructors often have students write — laboratory reports, term papers, essays, summaries, critiques, research proposals — but generally lack the time (and perhaps the confidence) to discuss the writing in much detail and with much conviction. Nevertheless, instructors know that written communication is an important part of the biologist's trade; most of us wish we had more time to teach our students to do it well. Moreover, since bad scientific writing often reflects fuzzy thinking, questioning the writing generally guides students toward a clearer understanding of the biology being written about. There are two good reasons, then, to promote better writing among students in biology courses: to give students training in a key professional skill, and to foster increased understanding of the subject matter. To improve students' writing in biology, half the battle is to persuade them to take the task seriously; the other half is to provide sufficient instructions to get them through the struggle successfully. In this book I address both halves of the battle, in a way that should not take up valuable class time.

The book is brief enough to be read by students along with other, more standard assignments, and straightforward enough to be understood without additional instruction. Although intended primarily for undergraduate use in typical lecture and laboratory courses, it is also appropriate for undergraduate and graduate seminars.

The benefits of learning to write well in Biology are discussed in Chapter 1, where I also describe the sort of writing that professional biologists do and review the key principles that characterize all sound scientific writing. This chapter also includes a section on the use of computers in preparing writing assignments. Chapter 8 is focused on the process of revision and includes some exercises in proofreading and editing. Chapters 1 and 8 should be read by all students early in the term before beginning any writing assignments; the other chapters cover the specific forms of writing typically encountered in the biology undergraduate curriculum and can be assigned as appropriate, in any order. Chapter 2, Writing

Laboratory Reports, is the longest in the book, and emphasizes that the result obtained in a study is often less important than the ability to discuss and interpret that result convincingly in the context of basic biological knowledge, and to demonstrate clear understanding of the purpose of the study. I emphasize the variability inherent in biological systems and how that variability is dealt with in presenting, interpreting, and discussing data. This chapter should also be useful for graduate students preparing papers for publication. Chapter 3, Writing Essays and Term Papers, discusses the most profitable ways to decide on and explore a paper or essay topic. It also includes special sections on the art of effective notetaking and organization, and the use of specialized indexing services such as *Biological Abstracts*.

Writing research proposals, critiques and summaries, and in-class essay examinations are discussed in separate chapters (Chapters 4, 5, and 9). To make the book more useful to all Biology majors, I have included chapters on how to give oral presentations (Chapter 6) and how to prepare applications for summer and permanent jobs in Biology and for graduate and professional schools (Chapter 7). There is also an appendix listing commonly used abbreviations for lengths, weights, volumes, and concentrations.

I am happy to thank Professors Edward H. Burtt (Ohio Wesleyan University) and Carl Schaefer (University of Connecticut) for their comments and suggestions on a draft of the first several chapters, and am especially grateful to Professors Sylvan Barnet (Tufts University, retired) and Marcia Stubbs (Wellesley College) for reading several drafts of each chapter and kindly and patiently helping me to follow my own advice. Dawn Terkla (Tufts University) made several very helpful suggestions regarding the section on statistical analysis of data. My wife, Lindy, assisted in all aspects of manuscript preparation with good humor and provided encouragement and criticism at all crucial junctures. I also wish to thank everyone at Little, Brown and Company who helped bring this book to completion, especially Carolyn Potts and Barbara Breese.

I welcome comments from all readers; I have learned much about writing in preparing this book and hope that reading it will prove equally beneficial to others, and perhaps enjoyable as well.

Contents

They scribbled on their pads and scratched out what they had scribbled. Some didn't write anything. Some crumpled their paper. They began, in fact, to look like writers. An awful silence hung over the room, broken only by the crossing out of sentences and the crumbling of paper. They began to sound like writers.

On Writing Well, W.K. Zinsser. (1980).

Write to illuminate, not to confuse.

Attributed to the biologist Naj A. Kînehcép.

1
Introduction and General Rules

The logical development of ideas and the clear, precise, and succinct communication of those ideas through writing are among the most difficult, but most important, skills that can be mastered in college. Effective writing is also one of the most difficult skills to teach. This is especially true in Biology classes, where there is often much writing to be done but little time to focus on doing it well. The chief message of this book is that developing your writing skills is worth every bit of effort it takes, and that Biology is a splendid field in which to pursue this goal.

WHAT DO BIOLOGISTS WRITE ABOUT, AND WHY?

The sort of writing that biologists do (lectures, letters of recommendation, grant proposals, research papers, and critiques of research papers and grant proposals written by other researchers) is similar in many respects to the sort of writing (essays, literature reviews, term papers, and laboratory reports) you are asked to do while enrolled in a typical Biology course. Basically, we must all prepare arguments.

Like a good term paper or essay, a lecture is an argument; it presents information in an orderly manner and seeks to convince the audience that this information fits sensibly into some much larger story. Few students are aware of the time and effort required to write a coherent lecture, but the sad fact is that putting together a string of three or four lectures and then moving on to the next topic is the equivalent of preparing one twenty- to thirty-page term paper weekly.

In addition to preparing lectures, many of us spend quite a bit of time writing grant proposals, in the hope of obtaining the funding that will enable us to pursue our research programs (and possibly hire one of you for a summer in the process). A research proposal is unquestionably an argument; success depends on our being able to convince a panel of other biologists that what we wish to do is worth doing, that we are capable of doing it, and that it cannot be done without the funds requested. Research money is not plentiful. Even well-written proposals have a difficult time; poorly written proposals generally don't stand a chance.

When we are not writing grant proposals or lectures, we are often preparing the results of our research for publication. Essentially, these articles are laboratory reports based on data collected over a much longer period of time than the typical laboratory session; in research articles, as in the preparation of laboratory reports, the goal is to present data clearly, and to interpret those data thoroughly and convincingly in the context of other work and basic biological principles. The preparation of research reports typically involves the following steps:

> organization of the data;
>
> preparation of a first draft of the article (following the procedures outlined in Chapter 2 of this book);
>
> revision and retyping of the paper;
>
> critical review of the work by one or several colleagues;
>
> revision of the paper in accordance with the comments and suggestions of the readers;
>
> retyping and proofreading of the paper;
>
> and, finally, sending the paper to the editor of the journal in which we would most like to see our work published.

This is not the end of the story. The editor then sends the manuscript out to be reviewed by two or three other biologists. Their comments, along with those of the editor, are then sent to the author, who must again rewrite the paper, often extensively. The editor may then accept the revised manuscript, or may request that it be rewritten again prior to publication.

Biologists obviously write about Biology, but they also write about other things. One of the other things college and university biologists write about is you; letters of recommendation are especially troublesome for us because they are so important to you. Like a good laboratory report, literature review, essay, or term paper, a letter of recommendation must be written clearly, developed logically, and proofread carefully if it is to argue convincingly on your behalf and help get you where you want to go.

And then there are the progress reports, committee reports, and internal memoranda. All this writing involves thinking, organizing, nailing down a convincing argument on paper, revising, retyping, and proofreading.

I hope you are now convinced that effective writing is not irrelevant in a scientific career. When students in a Biology course receive criticisms of their writing, they often complain that "this is not an English course." These students do not understand that clear, concise, logical writing is an important tool of the biologist's trade, and that learning how to write well is at least as important as learning how to use a balance, extract a protein, use a taxonomic key, measure a nerve impulse, or run an electrophoretic gel. And, unlike these rather specialized techniques, mastering the art of effective writing will reward you regardless of the field in which you eventually find yourself. The fact that you may not become a biologist is no reason to cheat yourself out of the opportunity to become an effective writer; the difference between a well-crafted and poorly crafted letter of application is often the difference between getting the job you want or losing it to another contender.

THE KEYS TO SUCCESS

There is no easy way to learn to write well in Biology. All good writing involves two struggles: the struggle for understanding and the struggle to communicate that understanding to a reader. Like the making of omelettes or crepes, the skill improves with practice. However, being aware of certain key principles will ease the way considerably. Each of the following rules is discussed more fully in later chapters.

1. Work to understand your sources. When writing laboratory reports, spend time wrestling with your data until you are convinced you see the significance of what you have done. When taking notes, reread sentences you don't understand and look up the meaning of words that puzzle you. Too few students take this struggle for understanding seriously enough, but all good scientific writing begins with this struggle. You can excel — in college and in life after college — by being one of the few who meet this challenge head on. If you don't commit yourself to winning this first struggle, you will end up with nothing to say, or worse, what you do say will be wrong. In both cases you will produce nothing worth reading.

2. Think about where you are going before you begin to write. Much of the real work of writing is in the thinking that must precede each draft. Effective writing is like effective sailing; you must take the time to plot your course before getting too far from port. Your ideas about where you are going and how best to get there may very well change as you continue to work with and revise your paper, since the act of writing invariably clarifies your thinking and often brings entirely new ideas into focus. Nevertheless, you must have some plan in mind even when you first begin to write. This plan evolves from thoughtful consideration of your notes. Think first, then write; thoughtful revision follows. If, when you sit down to write that last draft of your paper, you still don't know where you are heading, you certainly won't get there smoothly, and you may well not get there at all. Almost certainly your reader will never get there.

3. Write to illuminate, not to confuse. Use the simplest words and the simplest phrasing consistent with that goal. Define all specialized terminology. In general, if a term was recently new to you, it should be defined in your writing. Don't try to impress the reader with big words and a technical vocabulary, but instead focus on getting your point across.

4. Make a statement and back it up. Remember, you are making an argument. In any argument, a statement of fact or an opinion becomes convincing to the critical reader only when that statement is supported by evidence or explanation; provide it. You might, for instance, write, "Among the vertebrates, the development of sperm is triggered by the release of the hormone testosterone (Browder, 1980)." In this case, the statement is supported by reference to a book written by Browder in 1980. In the following example, a statement is backed up by reference to the writer's own data: "Some wavelengths of light were more effective than others in promoting photosynthesis. For example, the rate of oxygen production at 650 nm[1] was nearly four times greater than that recorded for the same plants when using a wavelength of 550 nm (Figure 2)."

5. Always distinguish fact from possibility. In the course of examining your data or reading your notes, you may form an opinion. This is splendid. But you must be careful not to state your opinion as though it were fact. "Species X lacks the ability to respond to sucrose" is a statement of fact, and should be supported with a reference. "Our data suggest that species X lacks the ability to respond to sucrose" or "Species X seems unable to respond to sucrose" expresses your opinion and should be supported by drawing the reader's attention to key elements of your data set.

6. Say exactly what you mean. Words are tricky; if they don't end up in the right places, they can add considerable ambiguity to your sentences. "I saw three squid SCUBA diving last Thursday" conjures up a very interesting image. Don't make the readers guess what you're trying to say; they often guess incorrectly. Good scientific writing is precise. Write to mean what you mean to say, and be sure you say what you mean.

7. Never make the reader back up. You should try to take the reader by the nose in your first paragraph and lead him or her through to the end, line by line, paragraph by paragraph. Avoid making the reader flip back two pages, or even one sentence. Link your sentences carefully, using such transitional words as "Therefore"

---

[1] nm = nanometers; i.e., 10^{-9} meters

or "In contrast," or by repeating key words, so that a clear argument is developed logically. Remind the reader of what has come before, as in the following example.

> In saturated air (100% relative humidity), the worms lost about 20% of their initial body weight during the first 20 hours but were then able to prevent further dehydration. In contrast, worms maintained in air of 70–80% relative humidity experienced a much faster and continuous rate of dehydration, losing 63% of their total body water content in 24 hours. As a consequence of this rapid dehydration, most worms died within the 24–hour period.

Note that the second and third sentences in this example begin with transitions ("In contrast, . . .", "As a consequence of . . ."), thus continuing and developing the thought initiated in the preceding sentences. A far less satisfactory last sentence might read, "Most of these animals died within the 24-hour period."

Link your paragraphs in the same way, using transitions to continue the progression of a thought, reminding the readers periodically of what they have already read.

Avoid casual use of the words *it, they,* and *their.* For example, the sentence "It can be altered by several environmental factors" forces the reader to go back to the preceding sentence, or perhaps even to the previous paragraph, to find out what *it* is. Changing the sentence to "The rate of population growth can be altered by several environmental factors" solves the problem. Here is another example:

> Our results were based upon observations of short–term changes in behavior. They showed that feeding rates did not vary with the size of the caterpillar.

The word *they* could refer to "results," "observations," or "behavior." Granted, the reader can back up and figure out what *they* are, but you should work to avoid the "You know what I mean" syndrome. Changing *they* in the second sentence to "These results" avoids the ambiguity and keeps the reader moving in the right direction.

Do not be afraid to repeat a word used in a preceding sentence; if it is the right word and avoids ambiguity, use it.

8. Be concise. Give all the necessary information, but avoid using more words than you need for the job at hand. By being concise, your writing will gain in clarity. Why say,

```
Our results were based upon observations of short-
term changes in behavior. These results showed that
feeding rates did not vary with the size of the
caterpillar.
```

when you can say the following?

```
Our observations of short-term changes in behavior
indicate that feeding rates did not vary with the size of
the caterpillar.
```

In fact, you might be even better off with the following sentence:

```
Feeding rates did not appear to vary with the size of
the caterpillar.
```

With this modified sentence, fifty percent of the words in the first effort have been eliminated without any loss of content. The savings are not merely esthetic. It costs something like twenty cents a word to publish a scientific paper, and authors are often asked to bear some of this cost; in the real world of biological publications, it pays, quite literally, to be concise. Besides, cutting out extra words means you will have less to type. You'll have your paper finished that much sooner. Finally, your reader can digest the paper more easily, reading it with pleasure rather than with impatience.

9. Stick to the point. Delete any irrelevant information, no matter how interesting it is to you. Snip it out and put it away in a safe place for use in later life if you wish, but don't let asides interrupt the flow of your writing.

10. Write for your classmates and for your future self. It is difficult to write effectively unless you have a suitable audience in mind. It helps to write papers that you can imagine being understood by your fellow students. You should also prepare your assignments so that they will be meaningful to *you* should you read them far in the future, long after you have forgotten the details of coursework completed or experiments performed. Addressing these two audiences — your fellow students and yourself — should help you write clearly and convincingly.

11. Don't plagiarize. Express your own thoughts in your own words. If you are quoting from another writer, you must credit your source explicitly. Note, too, that simply changing a few words here and there, or changing the order of a few words in a sentence or paragraph, is still plagiarism. It can be a dangerous game to play.

12. Appearances can be deceiving. Type your papers if possible, using only one side of each page. Leave margins of about an inch and a half on the left and right sides of the page, leave about an inch at the top and bottom of each page, and double-space your typing so that your instructor can easily make comments on your paper. Make corrections neatly. Never underestimate the subjective element in grading.

13. Always underline species names: for example, <u>Homo</u> <u>sapiens</u>. Note also that the generic name (<u>Homo</u>) is capitalized whereas the specific name (<u>sapiens</u>) is not. Once you have given the full name of the organism in your paper, the generic name can be abbreviated; <u>Homo</u> <u>sapiens</u>, for example, becomes <u>H</u>. <u>sapiens</u>. There is no other acceptable way to abbreviate species names. In

particular, it is not permissible to simply refer to an animal using the generic name, since most genera include many species.

14. Please remember that the word *data* is plural. The singular is *datum*, a word rarely used in biological writing. "The data are lovely" (not "The data is lovely"). "These data show some very strange trends" (not "This data shows some interesting trends"). You would not say "My feet is very large"; treat *data* with the same respect. Leave misuse of that word to the computer people.

15. Don't be teleological. That is, don't attribute a sense of purpose to other living things, especially when discussing evolution. Giraffes did not evolve long necks "in order to reach the leaves of tall trees." Snails did not evolve shells "in order to confound predators." Birds did not evolve nest-building behavior "in order to protect their young." Insects did not evolve wings "in order to fly." Plants did not evolve flowers "in order to attract bees for pollination." Evolution proceeds through a process of differential survival and reproduction, not with intent. Long necks, hard shells, complex behavior, and other such genetically determined characteristics may well have given some organisms an advantage in surviving and reproducing unavailable to individuals lacking these traits, but this does not mean that any of these characteristics were deliberately evolved in order to achieve something.

Organisms cannot evolve structures, physiological adaptations, or behavior out of desire. Appropriate genetic combinations must always arise by random genetic events, by chance, before selection can operate. Even then, selection is imposed on the individual by its surroundings and, in that sense, selection is a passive process; natural selection never involves conscious, deliberate choice. Don't write, "Insects may have evolved flight in order to escape predators." Instead, write, "Flight in insects may have been selected for in response to predation pressure." Don't write, "The parent gulls remove the white, conspicuous eggshells in order to protect the newly hatched, black-headed young." Instead, write, "Parental removal of the white, conspicuous eggshells may protect the newly hatched, black-headed young from predation."

16. Proofread. Although it is an important part of the writing process, none of us likes to proofread. By the time we have arrived at this point, we have put in a considerable amount of work and

are certain we have done the job correctly. Who wants to read the paper yet another time? Moreover, finding an error means having to make a correction. But put yourself in the position of your instructor. Your instructor must read perhaps a hundred or more papers each term. He or she starts off on your side, wanting to see you earn a good grade. Similarly, a reviewer or editor of scientific research manuscripts starts off by wanting to see the paper under consideration get published. A sloppy paper — for example, one with many typographical errors — can lose you a considerable amount of good will as a student and later as a practicing scientist. For one thing, it's insulting. Failure to proofread your paper and to make the required corrections implies that you don't value the reader's time; that is not a flattering message to send, nor is it a particularly wise one. Never forget: There is often a subjective element to grading and to decisions about the fate of manuscripts and grant proposals. Furthermore, sloppy work may suggest to the reader that you take little pride in your own efforts. Lastly, a carelessly proofread paper may suggest to the reader that the research itself was carelessly performed. For all these reasons, shoddily prepared material can easily lower a grade, reduce the chance for acceptance of a manuscript or grant proposal, damage the writer's credibility, or cost an applicant a job or admission to professional or graduate school. Why put yourself in such jeopardy for a mere half-hour saved?

17. Always make a copy of your paper before submitting the original to your instructor. Even instructors sometimes lose things.

ON USING COMPUTERS IN WRITING

Advertisers insist that ownership of a computer will take the work out of your achievement. Armed with a personal computer, they say, you will see your spelling improve, your sentences make sense, your paragraphs become well organized, your ideas seem brilliant, and your grades soar. As the ever-optimistic reader of many computer-printed laboratory reports, I must inform you that the amazing transformation has yet to occur. Word-processing programs unfortunately fail to make the process of writing much easier, largely because they do not remove from the author the responsibility for thinking, organizing, revising, and proofreading.

Computers cannot think, organize, or revise for you, and, at least for biologists, they are of little help in proofreading. Several computer programs designed to catch misspellings are currently available for purchase, but although these may be helpful in some fields they do not seem especially useful in Biology. Biology is a field with much specialized terminology, most of which is of no use to non-biologists; these terms, therefore, do not find their way into the dictionaries that accompany computer spelling programs. Although you can easily add words to the computer's dictionary, the terminology in your papers will be changing with every new assignment; most of the words you add for today's assignment will not be used in next week's assignment. Why not just buy a good dictionary and use it? Moreover, a spelling checker program will not distinguish between *to* and *too*, *there* and *their*, or *it's* and *its*, and the program will miss typographical errors that are real words; using the program will not spare you the chore of proof-reading for spelling mistakes. Suppose, for example, that you typed *an* when you intended to type *and*, or you typed *or* when you should have typed *of*. To catch these errors you would have to run a program specializing in catching grammatical mistakes. It is difficult to see how you will save much time in Biology by using a computer to proofread for spelling or grammatical errors.

On the other hand, computers are a writer's best friend when it comes to making revisions on advanced drafts of manuscripts, and it is here you can exploit those disk drives to best advantage. Back in the days of typewriters, when reading drafts of my papers, I would often see places where the addition of a phrase or sentence, or even the simple replacement of one word by another, would substantially improve the final product. But if I had already typed several drafts, I rarely made these additional changes; the benefits of increased clarity of expression were usually outvoted by the unbearable thought of retyping one or more pages yet again. With a word processor, however, perfection is within your immediate grasp. It is now easy to change a word, modify or delete a sentence, or reorganize a paragraph, and the computer will print out the revised version at the touch of a button. But beware: word processing is a two-edged sword; because revisions no longer require time-consuming retyping, use of a computer places increased responsibility on the writer to see that the revisions get made. Instructors find it increasingly annoying when students turn in computer-printed reports that are carelessly written and not proofread.

First drafts serve primarily to get ideas on paper, where they can't escape; the form, order, and manner of expression are not major concerns at this early stage of the creation. Consequently, the first revision is often so extensive that it is far less time-consuming to revise this draft by hand than to word-process one's way through it. In fact, putting a first draft on the computer might actually inhibit you from making the extensive revisions that are called for. Moreover, when there is a power failure (when the chips are down?) or when your disk gets mangled in the midst of your brilliant insights, you can lose everything; I will still have my first, handwritten draft. For me, it's the second draft that gets entered into the computer.

If you are one of those people who are intimidated by a blank sheet of paper but not by a blank video screen, ignore my advice about not using a computer for first drafts. Otherwise, reserve the computer for a later version of your assignment.

In addition to their use as word processors, computers are also used by some biologists for data storage or analysis and for statistical evaluation of especially large or complex data sets. But undergraduate Biology majors will probably find that a thirty-dollar scientific calculator will be perfectly adequate for anything they will encounter in the laboratory.

If you have never used a personal computer for word processing, do not be intimidated by the jargon used by those who have. Getting started is easy; there are only a few basic steps and these apply to all computers and all word processing programs:

    a) Turn on the machine;
    b) Insert a blank floppy disk;
    c) Tell the computer to format the blank disc so that the information you put on it can be reread by the machine later;

d) Activate the word processing program;
e) Open a file in which you will store your soon-to-blossom document;
f) Type your document;
g) Save what you have typed so that it can be called forth at will in precisely the form in which you left it;
h) Print your document.

Learning to edit with a computer can be confusing, but only if you try to learn everything at once. Beginning editors need to learn only these few basic commands to get the job done: move left within a line, move right within a line, move up a line, move down a line, delete a letter or space, add a new letter or space, reformat a paragraph (tidy up) after adding or deleting material. After you gain confidence using these basic operations, you can learn trickier maneuvers (such as deleting entire words or sentences with a single punch of a button, or moving sentences and paragraphs from one place to another) at your leisure.

Summary

1. Acknowledge the struggle for understanding and work to emerge victorious; read with a critical, questioning eye.
2. Think about where you are going before you begin to write, while you write, and while you revise.
3. Write to illuminate, not to confuse.
4. Back up all statements of fact or opinion.
5. Always distinguish fact from possibility.
6. Say exactly what you mean.
7. Never make the reader back up.
8. Be concise.
9. Stick to the point.
10. Write for an appropriate audience: your classmates and your future self.
11. Don't plagiarize.
12. Make your papers neat in appearance.
13. Underline the scientific names of species.

14. "Data are . . ."
15. Avoid teleology.
16. Proofread all work before turning it in, and keep a copy for yourself.
17. When possible, use word processing computer programs for revising advanced drafts of manuscripts, but prepare the first draft or two using pen, pencil, or typewriter.

2

Writing Laboratory Reports

WHY ARE YOU DOING THIS?

It is no accident that most Biology courses include laboratory components in addition to lecture sessions. Doing Biology involves asking questions, formulating hypotheses, devising experiments to test the hypotheses, presenting data, evaluating data, and interpreting data. Those so-called facts you learn from lectures and textbooks are primarily interpretations of data. By participating in the acquisition and interpretation of data, you glimpse the true nature of the scientific process.

If you are contemplating a career in research, be assured that learning to write effective laboratory reports now is an investment in your future. As a laboratory technician or research assistant, you will often be asked to work up and graph data so that the future path of the research can be decided upon. If you eventually pursue a research master's degree or Ph.D., you will find that a graduate thesis is essentially a large lab report. Writing up your research for publication, as a graduate student or as a researcher with a laboratory of your own, you will quickly find that you are again following the procedures you used in preparing good laboratory reports in college Biology courses; learn the tools of your trade now.

Preparing laboratory reports develops the ability to organize ideas logically, think clearly, and express yourself accurately and concisely. It is difficult to imagine a career in which mastery of such skills is not a great asset.

COMPONENTS OF THE LABORATORY REPORT

A laboratory report is typically divided into five major sections:

1. *Introduction.* The introductory section, usually only one or two paragraphs long, tells why the study was undertaken; a brief summary of relevant background facts leads to a statement of the specific problem that is being addressed.

2. *Materials and Methods.* This section is *your* reminder of what you did, and it also serves as a set of instructions for anyone wishing to repeat your study in the future.

3. *Results.* What were the major findings of the study? Present the data or summarize your observations, using graphs and tables so as to reveal any trends you found. Point out these trends to the reader. If you make good use of your tables and graphs, the results can commonly be presented in only one or two paragraphs of text; one picture is worth quite a few words. Avoid interpreting the data in this section.

4. *Discussion.* How do your results relate to the goals of the study, as stated in your introduction, and how do they relate to the results that might have been expected from background information obtained in lectures, textbooks, or outside reading? What new hypotheses might now be formulated, and how might these hypotheses be tested? This section is typically the longest part of the report.

5. *Literature Cited.* This section includes the full citations for any references (including textbooks and laboratory handouts) you have cited in your report. Double-check your sources to be certain they are listed correctly, because this list of citations will permit the interested reader to check the accuracy of any factual statements you make, and often, to understand the basis for your interpretations of the data. Cite only material you have actually read.

In some cases you will be asked to include an additional section in your report, the Abstract, in which you summarize the nature of the problem addressed and the major findings and conclusions.

Before writing your first report, it is helpful to study a few short papers in a major biological journal, such as *Biological Bulletin* or *Ecology*. Reading these journal articles for content is unnecessary; you don't need to understand the topic of a paper to appreciate how the article is crafted. But do pay attention to the way the Introduction is constructed, the amount of detail included in the Materials and Methods section, and the material that is and is not included in the Results section.

In studying an article or two, note that figures and tables are always accompanied by explanatory captions, and that the axes of graphs and the columns and rows of tables are clearly labelled.

Where to Start

Strangely enough, the Introduction is not the place to begin writing your report; it is far easier to write the Introduction towards the end of the job, after you have fully digested what it is that you have done. Start work with either the Materials and Methods section or with the Results section. Better still, you may profitably work on the two in tandem; working on the Results section sometimes helps to clarify what should be included in the Methods section, and working on the Methods section sometimes clarifies the order in which results should be presented in the Results section.

Because the Materials and Methods section requires the least mental effort, completing it is a good way to overcome inertia. You may not know why you did the experiment or what you found out by doing the experiment, but you can probably reconstruct what you did without much difficulty. Moreover, reminiscing carefully about what you did puts you in the right frame of mind to consider why you did it.

AN ASIDE ON CITING SOURCES

As stated earlier, every fact must be supported with a reference to its source. Here are a few general rules to follow when backing up factual statements. These rules apply to all sections of your report.

1. Don't footnote. In most papers published in biological journals, references are cited directly in the text, by author and year of publication, as in the following example:

> Many marine gastropod species enclose their fertilized
> eggs within structurally and chemically complex encapsu-
> lating structures (Hunt, 1966; Tamarin and Carriker,
> 1968).

When more than two authors have collaborated on a single publication, a shortcut is permissible:

> A mutation is defined as any change occurring in the ni-
> trogenous base sequence of DNA (Tortora et al., 1982).

The *et al.* is an abbreviation for *et alii,* meaning "and others." The words are underlined, even when abbreviated, because they are in a foreign language, Latin; underlining tells a printer to set the designated words or letters in italics. Note that in each of the examples given, the period follows the closing parenthesis, since the reference, including the publication date, is part of the sentence. Where appropriate, you may incorporate the authors' names directly into a sentence:

> The phenomenon of bioluminescence has been carefully re-
> viewed by Nicol (1967).

or

> Nicol (1967) has carefully reviewed the literature on
> bioluminescence.

se in citing references. Avoid writing,

sic work, The Biology of Marine Animals, pub-
967, Colin Nicol reviewed the literature on
;e bioluminescence.

non of invertebrate bioluminescence has been
eviewed by Nicol (1967).

Again, the period follows the parenthesis.

3. Cite only those sources you have actually read. Don't list references simply to add bulk to this section of your report; your instructor is perfectly justified in expecting you to be able to discuss any material you cite. Listing a few references you have thoughtfully incorporated into your paper should do more for your grade than any attempt to create the illusion that you have read everything in the library.

You may occasionally have to cite a reference that you have not actually read. For example, the results reported by Smith (1964) may be cited in a book or article written by Jones (1983), and you have read only the work by Jones. Your citation should then read, "(Smith, 1964; as cited by Jones, 1983)." Let Jones take the blame if he or she has misinterpreted something.

WRITING THE MATERIALS AND METHODS SECTION

Results are meaningful in science only if they can be obtained over and over again, whenever the experiment is repeated. Unfortunately, the results of any study depend to a large extent on the way the study was done. It is therefore essential that you describe your methodology in detail sufficient to permit your experiment to be repeated exactly as originally performed. Perhaps the best reason for writing a detailed Materials and Methods section is that it helps you to review what you have done in an organized way and starts you thinking about why you've done it. Developing a good Materials and Methods section puts you in the right frame of mind to do an equally good job on the other sections of the report.

The difficulty in writing this section of a laboratory report (or journal manuscript) is in selecting the right level of detail. Students commonly give too little information; when informed of this defect, they may then give too much information. It's hard to hit it just right, but keeping your audience in mind (yourself and your fellow students) will help.

Many students begin with a one-sentence Materials and Methods section: "Methods were as described in the lab manual." Although this sentence meets the criterion of brevity, it is generally unacceptable as a stand-alone Methods section. For one thing, studies are rarely performed exactly as described in a laboratory manual or handout. Your instructions may call for the use of 15 animals, for example, but only 12 animals might be available for use on the day of your experiment. In addition, many details of a study will vary from year to year, week to week, or place to place, and must therefore be omitted from your set of instructions.

But don't get carried away! Consider the following overly detailed description of a study involving the growth of radish seedlings:

> On January 5, I obtained four paper cups, 400 g of potting soil, and 12 radish seeds. I labelled the cups A, B, C, D and planted three seeds per cup, using a plastic spoon to cover each seed with about one-quarter inch of soil.

The author has used the first sentence simply to list the materials; whenever possible, it is far better to mention each new material as you discuss what you did with it. Also, why do we need to know the weight of soil obtained, or that the cups were labelled

A–D rather than 1–4, or that a plastic spoon was used to add soil? Omitting the excess details and starting right in with what was done, we obtain:

```
On January 5, I planted three radish seeds in each of
four individually marked paper cups, covering the seeds
with about one-quarter inch of potting soil.
```

Note that the essential details — individually marked cups, three seeds per cup, one-quarter inch of soil — not only survive in the edited version, but stand out clearly. The trick, then, is to determine which details are essential and which are not.

The best approach to writing the Materials and Methods section is to begin by listing all the factors that might have influenced your results. If, for example, you measured the feeding rates of caterpillars on several different diets, your list might look something like this:

species of caterpillar used

diets used

amount of food provided per caterpillar

time of year

time of day

air temperature in room

manufacturer and model number of any specialized equipment used (such as balances, centrifuges, or spectrophotometers)

size and age of caterpillars

duration of the experiment

container size

number of animals per container

total number of individuals in the study

This list, which you do not turn in with your report, contains the bricks with which you will construct the Materials and Methods section. Each of the listed details must find its way into your report (not necessarily in the order in which you jotted them down) because each gives information essential for later replication of the experiment. Some of this information may also help you explain why your results differed from those of others who have gone before you, a topic that will deserve some attention later, in the Discussion section of your report. Details that do not merit inclusion in the list are superfluous and should not appear in your Materials and Methods section.

In describing the procedures followed, you must say what you did, but you should freely refer to your laboratory manual or handouts in describing how you did it. For example, you might write:

```
The three different diets were distributed to the cater-
pillars in random fashion, as described in the labora-
tory manual (Smith and Smith, 1983).
```

The important point here is that the diets were distributed at random; the outcome might be quite different if the largest caterpillars were to receive one diet and the smallest caterpillars another. The interested reader, including you, perhaps, at some later date, can refer to the stated source (Smith and Smith, 1983) for detailed instruction in the method of randomization. You might want to append the relevant portion of your handout or manual at the end of your report, as an appendix; this is a fine way to keep everything together for later use.

It is often a good idea to mention, for your own benefit as well as that of your reader, why particular steps were taken. Imagine yourself explaining things to a classmate. We might, for example, profitably rewrite the sentence given as an example in the preceding paragraph to read:

```
To avoid prejudicing the results by distributing food
according to size of caterpillar, the three different
diets were distributed to the caterpillars in random
fashion as described by Smith and Smith (1983).
```

It is also usually appropriate to include any formulas used in analyzing your data. The following sentences, for example, would belong in a Materials and Methods section:

```
The data were analyzed by a series of Chi-Square tests.
The rate at which food was eaten was calculated by divid-
ing the weight loss of the food by three hours, according
to the formula
```

```
Feeding rate = (Initial food weight - final food weight) ÷ 3 h.
```

Be sure to note any departures from the given instructions. Suppose you were told to weigh the caterpillars individually but found that your balance was not sensitive enough to record the weight of a single animal. Your laboratory instructor, never at a loss for good ideas, probably suggested that you weigh the individuals in each container as a group. Your report might then include the following information:

```
Determining the weight gained by each caterpillar
over the three-hour period of the experiment required
that both initial and final weights be determined. The
caterpillars were too small to be weighed individually.
Therefore, similarly sized caterpillars were weighed in
groups of three at a time. The average weight of each
caterpillar in the group was then calculated.
```

The Materials and Methods section of your report should be brief but informative. The following example completely describes an experiment designed to test the influence of decreased salinity on the body weight of a marine worm.

The polychaete worms used in this study were <u>Nereis</u> <u>virens</u>, freshly collected from Nahant, MA, and ranging in length between 10 and 12 cm. All treatments were performed at room temperature, approximately 21°C, on April 15, 1984. One hundred ml of full-strength seawater was added to each of three 200 ml glass jars; these jars served as controls, to monitor worm weight in the absence of any salinity change. Another three jars were filled with 100 ml of seawater diluted by 50% with distilled water.

Six polychaetes were quickly blotted with paper towels to remove adhering water, and were then weighed to the nearest 0.1 g using a Model MX-200 Fisher/Ainsworth balance. Each worm was then added to one of the jars of seawater. Blotted worm weights were later determined at 30, 60, and 120 minutes after the initial weights were taken.

The initial and final osmotic concentrations of all test solutions were determined using a Model 3W Advanced osmometer, following instructions provided in the handout.

Note that all essential details have been included: temperature, species used, size of animals used, number of animals used per treatment, number of animals per container, volume of fluid in the containers, type and size of containers, time of year, and equipment used. After reading this Materials and Methods section, you could repeat the study if you wanted to (or had to). Note, too, that the writer has made clear why certain steps were taken; three jars of full-strength seawater served as controls, for example, and worms were blotted dry to remove external water. The fact that worms were blotted dry before they were weighed was mentioned because this is a procedural detail that would obviously influence the results. On the other hand, the way the balance was operated was omitted, since this technique is standard. The author has written a report that might be useful to him or her in the future . . . and ends up with a top grade.

On to the Results.

WRITING THE RESULTS SECTION

In this section, you summarize your findings, using tables, graphs, and words. The Results section is

1. not the place to discuss why the experiment was performed;
2. not the place to discuss how the experiment was performed;
3. not the place to discuss whether the results were expected, unexpected, disappointing, or interesting.

Simply present the results. Don't interpret them.

Tables and Graphs

Before you even think about doing the writing part of your Results section, you must work with your data. The observations you've made, the data you've collected, most likely contain a story that is crying out for recognition. Contrary to popular opinion, the purpose of tabling and graphing data is not to add bulk to laboratory reports. You must manipulate the data in tables and graphs in order to reveal trends, not only to your instructor but, more important, to yourself. The trick now is to organize the data so that (1) the underlying story is revealed and (2) the task of revealing the story to your reader is simplified.

There is no single right way to present summaries of data; use whatever system gives the clearest illustration of trends. You must first decide what relationships might be worth examining, and then experiment with different ways of tabulating and graphing the data to best explore and demonstrate those relationships. Suppose we return to the experiment in which we measured the rates at which caterpillars fed for three hours on three different diets. We determined both the initial weight of food provided and the weight of food remaining after the three-hour period, so we can calculate the weight of food eaten per caterpillar per hour. In your report, you should provide a sample calculation, so that if you make a mistake your instructor can see where you went wrong. We also know the initial and final weights of the caterpillars for each diet, and the initial and final weights of dishes of food in the absence of caterpillars; these control dishes will tell us the amount of water lost by the food because of evaporation.

What relationships in the data might be especially worth examining? The first step in answering this question is to make a list of specific questions that might be worth asking:

1. Do the caterpillars feed at different rates on the different diets? That is, does feeding rate vary with diet?
2. Do larger caterpillars eat faster than smaller caterpillars? That is, does feeding rate vary with size of caterpillar?
3. How is the weight gained by a caterpillar related to the weight lost by the food?
4. Did the weight of the control dishes change, and, if so, by how much?

As in preparing the Materials and Methods section, this list is for your own use and is not to be included in your report. Don't take any shortcuts here. Write these questions in complete sentences. Once you have this list of questions, it is easy to list the relationships that must be examined in your Results section:

1. feeding rate as a function of diet;
2. feeding rate as a function of caterpillar size;
3. caterpillar weight gain versus food weight loss for each caterpillar;
4. food weight loss in the presence of caterpillars versus food weight loss in controls.

Now you must organize your data into a table in a way that will let you examine each of these relationships. Examine Table 1. This rough draft lists all the data obtained in the experiment; you can work from this table.

For the first relationship in our list (feeding rate as a function of diet), a table will tell the entire story. For your report, you can simply present a summary table, with explanatory caption, as in

Table 1. Summary of Raw Data

Diet	Initial Caterpillar Wt. (g)	Final Caterpillar Wt. (g)	Caterpillar Weight Change (g)	Food Wt. loss (g)	Feeding Rate (g food lost/h caterpillar)
A	8.05	9.55	+1.55	3.65	15.2×10^{-2}
A	4.80	5.80	+1.00	1.74	7.2×10^{-2}
A	5.50	7.00	+1.50	3.33	13.9×10^{-2}
A	5.50	4.70	~~-0.80~~	~~0.80~~	0
A	5.90	6.95	+1.05	1.35	5.6×10^{-2}
Average	5.95	6.80	+1.28	2.52	8.4×10^{-2}
B	4.40	5.11	+0.71	2.19	9.1×10^{-2}
B	5.20	5.60	+0.40	1.25	5.2×10^{-2}
.					
.					
Control 1	—	—	—	0.22	—
2	—	—	—	0.10	—
3	—	—	—	0.16	—

Table 2.   Average rates of food consumption over a 24 h period for caterpillars given three separate diets.

Diet	No. caterpillars	(g food eaten/caterpillar/h)
A	4*	8.4×10^{-2}
B	5	3.8×10^{-2}
C	5	7.9×10^{-2}

* One individual died during the study, without eating any food.

Table 2. Note that one of the caterpillars offered diet *A* ate no food and lost weight during the experiment. This individual died during the study, and the associated data were therefore omitted from Table 2. (The weight loss for this caterpillar probably reflects evaporation of body water.)

Finally, the time has come to reveal more subtle trends that may be lurking in the data. These trends may not be readily apparent from the summary table (Table 2); the trends may be made visible, however, to you and your reader, through graphing. A word of caution: Do not automatically assume that your data must be graphed. If you can tell your story clearly using only a table, a graph is superfluous.

Graphs in Biology generally take one of two basic forms: scatter plots (point graphs) or histograms (bar graphs). For the second relationship (feeding rate versus caterpillar size) we wish to examine using the caterpillar data, a scatter graph, like Figure 1, will be especially appropriate.

In examining Figure 1, please note that

1. each axis of the graph is clearly labelled;
2. the meaning of each symbol is clearly indicated;
3. a detailed explanatory caption ("figure legend") accompanies the figure.

All of your graphs should share these three characteristics. In the graph that the student labelled Figure 1, it would be insufficient to simply label the *Y* axis "Feeding Rate." Feeding rates can be

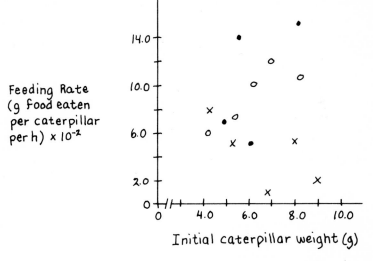

Figure 1.   The relationship between initial caterpillar weight and rates of food consumption for <u>Manduca</u> <u>sexta</u> feeding on three different diets.

expressed as per minute, per hour, per day, or per year, and can be expressed as per animal, per group of animals, or per gram of body weight. Similarly, it is unacceptable to label the *X*-axis as "Weight," or even as "Caterpillar Weight." Don't make the reader guess what you have done. From the figure caption, the axis labels, and the graph itself, the reader should be able to interpret the figure without reference to the text. Never make the reader back up; a good graph is self-contained.

The third relationship (animal weight gain versus food weight loss) might well be left in table form, since in this case the trend is readily discernible; caterpillars always gained less weight than that lost by the food. The same trend could be revealed more dramatically (or, let us say, more graphically) with a scatter plot, as shown in Figure 2, but a graph is not essential here. Again, note the steps taken to avoid ambiguity: the axes are labelled, units of measurement are indicated, symbols are interpreted on the graph, and the figure is accompanied by an explanatory legend (also called a figure caption). Note also that the symbols used in the student's Figure 2 are consistent with their usage in Figure 1. You should always use the same system of symbols throughout a report, so as not to confuse your reader; if filled circles are used to represent data obtained on diet *A* in one graph, filled circles should be used to represent data obtained on diet *A* in all other graphs.

The fourth relationship in our list considers food weight loss

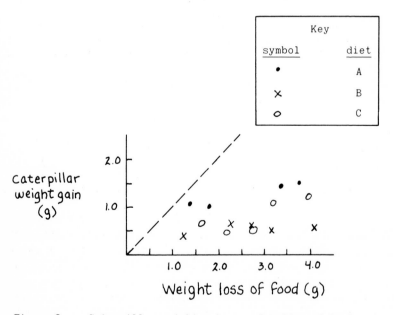

Figure 2.   Caterpillar weight gain as a function of food consumption for <u>Manduca</u> <u>sexta</u> fed on either of three diets. Points falling on the dotted line would indicate equality between weight gained and food eaten.

in control dishes (no caterpillars). No graphs or tables are needed here; two sentences will do:

> Control containers exhibited less than a 3% weight loss
> ($\underline{N}$ = 3 containers) during the 24 h period. In contrast,
> food in containers with caterpillars lost a minimum of
> 23% of initial weight.

If the weight loss had been substantial, perhaps 5–10% or more of initial weight, you might wish to adjust all the data in your tables accordingly, before making other calculations:

> Control containers exhibited a 7.6% weight loss ($\underline{N}$ = 3
> containers) over the 24 h period. Weight loss in other
> containers was therefore adjusted for this 7.6% evapora-
> tive loss before calculations of feeding rates were
> made.

You would then provide a sample calculation, so that your instructor could see how this was done, and so that you will remember what you did if you consult your report again at a later date. A less desirable but nevertheless acceptable alternative would be to state the magnitude of the evaporative weight loss in your Results section, and bring this point up again in interpreting your results in the Discussion section. In this case, you might want to label appropriate portions of graphs and tables as "Apparent Feeding Rates" rather than "Feeding Rates." Again, although there are many wrong ways to present the data, there is no single right way; you must simply be complete, logical, and clear.

So far, we have looked only at examples of tables and point plots. If you were studying the differences in species composition of insect populations trapped in the light fixtures on four different floors of your Biology building, a bar graph, as in Figure 3, might be more suitable. Note again that the axes are clearly labelled, including units of measurement, and that an explanatory legend

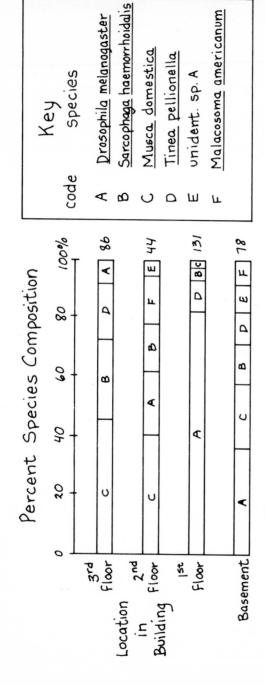

Figure 3. The distribution of insect species collected from light fixtures on four floors of the Biology building. The number to the right of each bar gives the total number of insects collected on each floor.

accompanies the figure. Don't make the reader back up. Note also that the graph tells an interesting story; given that *A* is the fruit fly *Drosophila melanogaster*, it is not difficult to guess where the genetics laboratory is located! Use tables and graphs only if they make your data work for you; if a table or graph fails to help you to summarize some trend in your results, it contributes nothing to your report and should be left out.

AN ASIDE ON TABLES AND GRAPHS

Tables should always be organized with like objects reading vertically rather than horizontally. Tables 3 and 4 present the same information, but in different formats. Table 3 correctly places all information about a single species in one row, so that readers can view the information for each species by scanning from left to right, and can compare data among different species by scanning up and down a single column. Table 4 is incorrectly organized and more difficult to read.

Graphs should always be constructed with the use of graph paper, which can be purchased in your campus bookstore. The most useful sort of graph paper has heavier lines at uniform intervals — at every ten divisions, for example, as shown in the graphs in Figures 4–8. These heavy lines facilitate the plotting of data and

Table 3.   Characteristics of four snail populations sampled at Nahant, MA on October 13, 1985.

Species	Average shell length (cm)	Sample size	Average no. animals per m^2
Crepidula fornicata	1.63	122 indiv.	32.1
C. plana	1.01	116	20.8
Littorina littorea	0.87	447	113.6
L. saxatilus	0.40	60	8.2

Table 4.   Characteristics of four snail populations sampled
at Nahant, MA on October 13, 1985.

Species	Crepidula fornicata	C. plana	Littorina littorea	L. saxatilus
av. shell length (cm)	1.63	1.01	0.87	0.40
sample size	122 indiv.	116	447	60
aver. no. animals per m^2	0.40	20.8	113.6	8.2

reduce eyestrain considerably, since every individual line need not
be counted in locating data points.

You need not draw directly on the graph paper; it should
be perfectly acceptable to place a piece of white tracing paper over
a piece of graph paper, trace the *X*- and *Y*-axes, mark off major
intervals, and plot points by seeing through the white paper to
the underlying grid. In this way, you can use one sheet of graph
paper many times. Taping the graph paper to your desk and taping
the tracing paper over it will prevent slippage and reduce frustration
(and errors). By convention, the independent variable is plotted
on the *X*-axis and the dependent variable is plotted along the *Y*-
axis. For example, if you examined feeding rates as a function of
temperature (Figure 4), you would plot temperature on the *X*-
axis and feeding rate on the *Y*-axis; feeding rate *depends* on temperature. On the other hand, temperature is not controlled by
feeding rate; that is, temperature varies independently of feeding
rate. Temperature is the independent variable and is plotted on
the *X*-axis.

It is good practice to label the axes of graphs beginning with
zero. To avoid generating graphs with lots of empty, wasted space,
breaks can be put in along one or both axes, as in Figure 4 (or
Figure 1 and Figures 7–9). If a break had not been inserted in the
Y-axis of Figure 4, for example, the graph would have been less
compact, as in Figure 5.

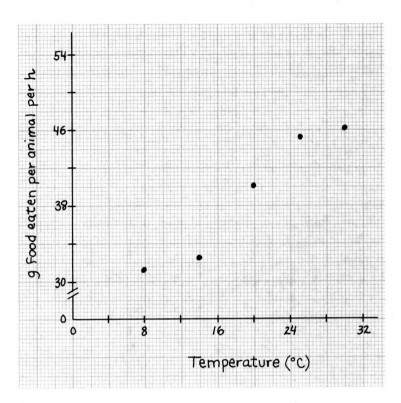

Figure 4.   Feeding rate of <u>Manduca</u> <u>sexta</u> caterpillars as a function of environmental temperature.

Connecting the Dots

After plotting data points, lines are often added to graphs to clarify trends in the data. It is especially important to add such lines if data from several different treatments are plotted on a single graph, as in Figure 6. Note that in this graph different symbols have been used for the data obtained at each temperature, to make the graph easier to interpret.

In some cases it makes more sense to draw smooth curves than to simply connect the dots. For example, suppose we have monitored the increase in height of tomato seedlings over a period of time in the laboratory. Every week we randomly selected 15–

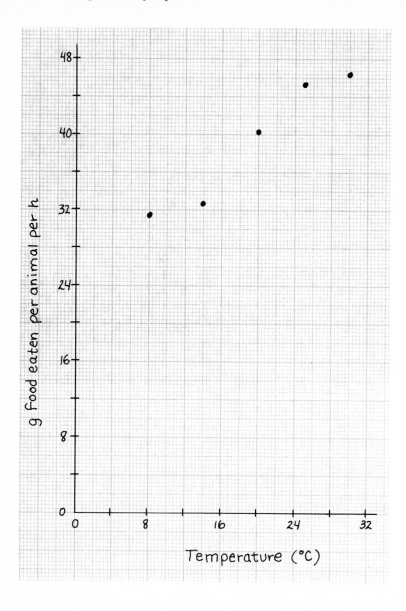

Figure 5.   Feeding rate of <u>Manduca</u> <u>sexta</u> caterpillars as a function of environmental temperature.

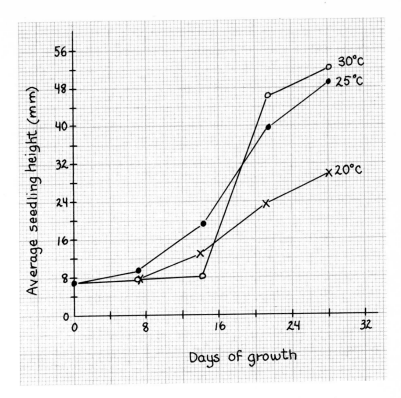

Figure 6.   Rate of seedling growth at three different temperatures.

20 seedlings to measure from the laboratory population of several hundred, so that different seedlings were usually measured at each sampling period. After two months, the data were plotted as in Figure 7.

Connecting the dots would not be the most sensible way to reveal trends in the data of Figure 7, since we know that the seedlings did not really shrink between days 21 and 28 and between days 35 and 42; simply connecting the points would suggest that shrinkage had occurred. This apparent decline in seedling height reflects the considerable variability in individual growth rates found within the sample population, as well as the fact that we did not measure every seedling in the population on every sampling day.

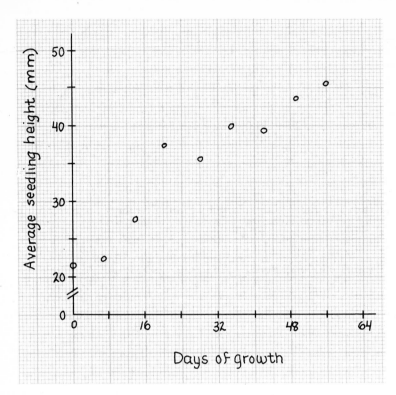

Figure 7.   Rate of seedling growth at 20°C. Fifteen—twenty
seedlings were measured on each day of sampling.

In this case, the trend in growth is best revealed by drawing a
smooth curve, as in Figure 8.

When plotting average values on a graph, it is appropriate
to include a graphic impression of the amount of variation present
in the data by adding bars extending vertically from each point
plotted (Figure 9). You may, for example, choose to simply illustrate
the range of values obtained in a given sample. In Figure 9, for
instance, the vertical bars extending from the point at day 30
indicate that although the average seedling height was about 37
mm (millimeters) on that day, at least one seedling in the sample
was as small as 25 mm, and at least one seedling was as large as
46 mm. Alternatively, you may plot the standard deviation or

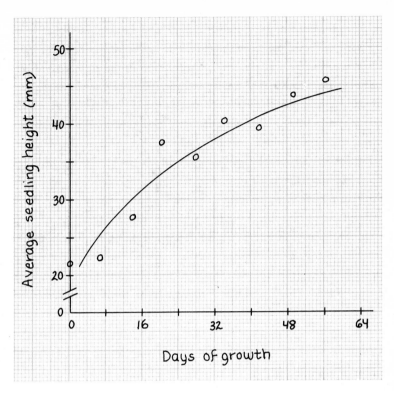

Figure 8.   Rate of seedling growth at 20°C. Fifteen—twenty
seedlings were measured on each day of sampling.

standard error about the mean, in which case the vertical lines will
extend equal distances above and below each point. Standard de-
viations and standard errors are reviewed in Appendix A (p. 178).
In your figure caption, be sure to indicate what it is you have
plotted.

When the variable along the X-axis (the independent variable)
is numerical and continuous, points can be plotted and trends can
be indicated by lines or curves as we have seen in Figures 1–9.
In Figure 4, for example, the X-axis shows temperature rising
continuously from 0°C to 32°C, with each cm along the X-axis
corresponding to a 4°C rise in temperature. Similarly, the X-axis

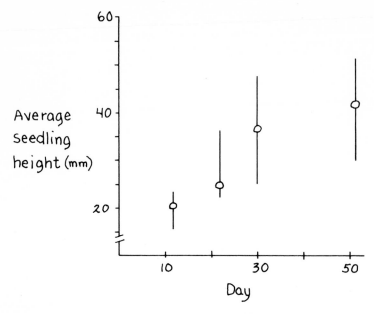

Figure 9.   Average height of tomato plant seedlings over a 50–day period. Vertical bars represent the range of heights.

of Figures 7–8 reflects the march of time, from 0 to 60 days or so, with each cm along the X-axis reflecting eight additional days.

   When the independent variable is non-numerical or discontinuous, or represents a range of measurements rather than a single measurement, a bar graph is appropriate, as shown in Figures 10 and 11. The X-axis of Figure 10 is labelled with the names of different vegetables. In contrast to the X-axes of Figures 1–9, the X-axis of Figure 10 does not represent a continuum; no particular quantity continually increases as one moves along the X-axis, and a line connecting the data for spinach and asparagus would be meaningless. In Figure 11, the data for shell length are numerical, but are grouped together (for example, all shells 25.0–29.9 mm in length are treated as a single data point). Note also that the magnitude of the size categories represented by the different bars

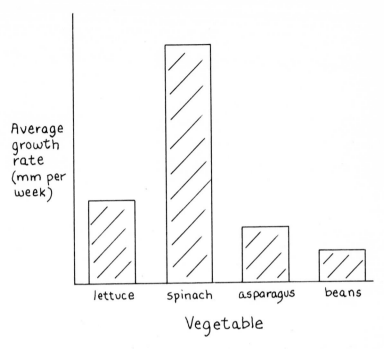

Figure 10.   Growth rates of four vegetables over a three-week period at 20°C.

varies; the left-most bar represents the percentage of shells found within a range of about 21 mm in length, whereas each of the next several bars to the right represents the percentage of shells found within a range of only about 5 mm in length. The size range of shells represented by the bar at the extreme right side of the graph is unknown; we know that all shells found in this category exceeded 45 mm in length, but the graph does not indicate the size of the largest shell.

One last thought about graphs, tables, and other forms of illustration: Be selective. Don't include a drawing or table unless you plan to discuss it, and include only those illustrations that best help you to tell your story.

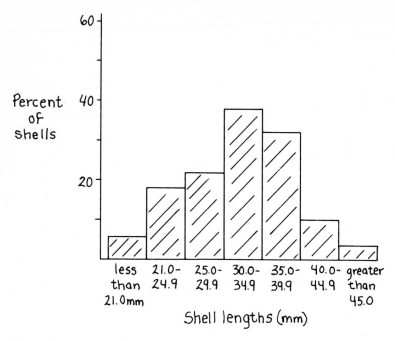

Figure 11.   Size distribution of snail shells (<u>Littorina</u>

<u>littorea</u>) collected at Blissful Beach, MA on August 15, 1985.

Only living animals were measured. A total of 197 snails were

included in the survey.

Verbalizing Results

One-sentence Results sections are common in student reports: "The results are shown in the following tables and graphs." However, *common* does not mean *acceptable*. You must use words to draw the reader's attention to the key patterns in your data. But do not simply redraw the graph in words, as in this description of Figure 6 (p. 37):

At 30°C, the seedlings showed negligible growth for

the first 10 days of study. However, between days 10 and

15, the average seedling grew nearly 15 mm, from about 8
mm to about 22 mm. Growth continued over the next 15 days,
with the seedlings reaching an average height of 45 mm by
day 20, and 52 mm by day 25.

Let the graph do this work for you; your task is to summarize
the most important trends displayed by the graph. For example,
you might write,

Temperature had a pronounced effect on seedling
growth rates (Figure 6). In particular, seedlings at 25°C
consistently grew more rapidly than those at 20°C . . .

Let us apply these principles to the caterpillar study. First,
is there anything about the general response of the animals worth
drawing attention to? You might, for example, be able to say:

All of the caterpillars were observed to eat throughout
the experiment.

More likely, living things behaving as they do, you will say something like:

One of the animals offered diet <u>A</u> and two of the animals
offered diet <u>B</u> were not observed to eat during the three-
hour experiment, and the results from these animals were
therefore excluded from analysis.

Such a decision to exclude data from further analysis is fine, by
the way, as long as you indicate the reason for the decision, and
as long as the decision is made objectively; you cannot exclude
data simply because it violates a trend that would otherwise be
apparent, or because the data contradict a favored hypothesis.

Next, go back to your initial list and reword each question as a statement. For example, the first question posed earlier, on p. 26 ("Do the caterpillars feed at different rates on the different diets?"), might be reworded as:

```
Caterpillars generally fed at faster rates on diet A than
on diet B.
```

In scientific writing, every statement of fact must be backed up with evidence. In this case you can support the statement with a simple reference to the appropriate figure or table:

```
Caterpillars generally fed at faster rates on diet A than
on diet B (Table 2).
```

Readers can then look at Table 2 and decide whether they see the same trend you saw. Sometimes you may want to give a specific example from your data as well, to further support your statement, but that is unnecessary here; one sentence and a picture say it all.

If you follow this procedure for each question on your list, your Results section will be complete. The written part will generally be fairly short.

Note that the statement about the rate at which caterpillars fed does not mention the term *significant*; it does not say, "Caterpillars fed at significantly faster rates on diet *A* than on diet *B*." Using the term *significant* implies that you have subjected the data to an appropriate statistical test to determine that the differences observed are substantial enough to be convincing. Do not write about significant differences among groups, or the lack of significant differences, unless you have conducted such a test. A brief introduction to the use of statistics in interpreting results begins on p. 68. This section is worth reading even if you are not required to conduct statistical analyses of your data; as biologists in training, the *how* of statistical analysis is less important to you than the *why*.

Note also the use of the past tense in the statement about caterpillar feeding rates:

```
Caterpillars generally fed at faster rates on diet A.
```

This statement is quite different from the following one, in which the present tense is used:

```
Caterpillars feed at faster rates on diet A.
```

By using the present tense, you would be making a broad generalization extending to all caterpillars, or at least to all caterpillars of the species tested. Before one can make such a broad statement, the experiment must be repeated many, many times, and similar results must be obtained each time; after all, the writer is making a statement about all caterpillars under all conditions. By sticking with the past tense here, you are clearly referring only to the results of your study. Be cautious.

Writing about Numbers

According to the Council of Biology Editors, you should use numerals when presenting percentages, decimals, magnifications, and abbreviated units of measurement (see pp. 180–181): 25%, 1.5 times greater, 15X magnification, 0.7 g, 18 ml. In most other situations, use words for numbers zero through nine, and numerals for larger whole numbers: seven seedlings, 15 flower petals, 247 protozoans. But all rules have exceptions. When using a series of numbers at least one of which is greater than 9, use numerals for all, as in the following example:

```
We added 6 drops to the first flask, 9 drops to the sec-
ond, and 12 drops to the third.
```

In Anticipation

Much of the work involved in putting together a good laboratory report goes into preparing the Results section. You can

Date and time started: _____

Date and time ended: _____

Caterpillar No.	Diet	Caterpillar wt.(g) Initial	Final	Weight change(g)	Food wt.(g) Initial	Final	Food wt. change(g)	Feeding rate g eaten/caterp/h
X	X	X	X		X	X		

Figure 12. Sample format for a laboratory data sheet.

save yourself considerable effort and frustration by planning ahead before you enter the laboratory to do the experiment. Be prepared to record your data in a format that will enable you to make your calculations easily. For the caterpillar experiment referred to previously, you would be well ahead by coming to the laboratory with a data sheet set up like the one in Figure 12. Using this data sheet, the data are recorded in the X areas during the laboratory period; the blank spaces will be filled in later, as you make your calculations. If possible, leave a few blank columns at the right, to accommodate unanticipated needs discovered as you record or work up your data. In introductory laboratory exercises, students are generally provided with data sheets already set up in a useful format. It is worthwhile to take a careful look at these data sheets in order to understand how they are organized and why they are as they appear; in more advanced laboratory courses, you will be responsible for organizing your own data sheets. One last suggestion: always follow any number you write down with the appropriate units, such as mg (milligrams) or cm (centimeters). This will avoid potential confusion later.

Following the advice of the previous paragraph can save you hours of work later.

Negative Results

An experiment that was correctly performed always "works." The results may not be what you had expected, but this does not mean that the experiment has been a waste of time. If biologists threw away their data every time something unexpected happened, we would rarely learn anything new. The data you collect are real; it is only the interpretation that is open to question. Therefore always treat your data with respect. The lack of a trend or the presence of a trend contrary to expectation is itself a story worth telling.

Putting together an effective Results section may seem like a lot of work. It is. But this section is the heart of your report; craft this section properly and the remainder of the work will be relatively easy.

WRITING A DISCUSSION SECTION

In this section of the report you must interpret your results in the context of the specific questions you set out to address in this experiment and in the context of any relevant broader issues that have been raised in lectures, textbook readings, previous coursework, and, possibly, library research. You will consider the following issues:

1. What did you expect to find, and why?
2. How did your results compare with those expected?
3. How might you explain any unexpected results?
4. How might you test these potential explanations?

Clearly, if your results coincide exactly with those expected from prior knowledge, your Discussion section will be rather short. Such a high level of agreement is rarely obtained in three- or four-hour laboratory studies. Indeed, high degrees of variability characterize many aspects of research in Biology, especially at the level of the whole organism. Often an experiment will need to be repeated many times, with very large sample sizes, before clear trends emerge. This point is discussed further in the section on statistical analysis beginning on p. 68. A short paper in a biological journal may well represent years of work by several competent, hardworking individuals. Even the simplest of questions are often not easily answered. Nevertheless, every experiment that was carried out properly tells you *something,* even if that something is not what you specifically set out to find out.

Expectations

State your expectations explicitly, and back your statements up with a reference. Scientific hypotheses are not simply random guesses. Your expectations must be based on facts, not opinions; these facts could come from lectures, laboratory manuals or hand-outs, textbooks, or any other traceable source. In discussing a study on the effectiveness of different wavelengths of light in promoting photosynthesis, for example, you might write something like:

All wavelengths of light are not equally effective in promoting photosynthesis; green light is said to be especially ineffective (Smith and Jones, 1981). This is because green light tends to be reflected rather than absorbed by plant pigments, which is why most plants look green (Smith and Jones, 1981). Our results supported this expectation. In particular . . .

Alternatively, a discussion section might profitably begin:

The results of our experiment failed to support the hypothesis (Smith and Jones, 1981) that caterpillars of <u>Manduca</u> <u>sexta</u> reared on one uniquely flavored diet will prefer that diet when subsequently given a choice of foods.

Here we have managed to state our expectations and compare them with our results in a single sentence. In both cases, we have begun our discussion on firm ground—with facts rather than opinions.

Explaining Unexpected Results

When results refuse to meet expectations, students commonly blame the equipment, the laboratory instructor, the laboratory partners, or themselves. Generally, more scientifically interesting possibilities than experimenter incompetence are the culprits. Don't be too hard on yourself. Take another look at the list of factors you wrote down when beginning to work on your Materials and Methods section. Could any of these factors be sufficiently different from the normal or standard conditions under which the experiment is performed to account for the difference in results? Look again at your laboratory manual or handout. Are any of the conditions under which your experiment was performed substantially different from those assumed in the instruction manual? If you discover no obvious differences in the experimental conditions, or if the differences cannot account for your results, include this point in your report:

> The discrepancy in results cannot be explained by the un-
> usually low temperature in the laboratory on the day of
> the experiment, since the control animals were subjected
> to the same conditions and yet behaved as expected.

If potentially important differences are noted, put this ammunition to good use:

> In prior years, these experiments have been performed
> using species $\underline{X}$. It is possible that species $\underline{Y}$ simply be-
> haves differently under the same experimental
> conditions.

Note that the writer does not *conclude* that species X and species Y behave differently; the writer merely *suggests* this explanation as a possibility. Always be careful to distinguish possibility from fact. Suggesting a logical possibility won't get you into any trouble. Stating your idea as though it were a proven fact, on the other hand, is sticking your neck out far enough to get your head chopped off.

Continue your discussion by indicating possible ways that the differences in behavioral responses might be tested. For example:

> This possibility can be examined by simultaneously ex-
> posing individuals of both species to the experimental
> conditions. If species $\underline{X}$ behaves as expected, and spe-
> cies $\underline{Y}$ behaves as it did in the present experiment, then
> the hypothesis of species-specific behavioral differ-
> ences will be supported. If species $\underline{X}$ and species $\underline{Y}$ both

```
respond as species Y did in the present study, then some
other explanation will be called for.
```

Continue in this vein, evaluating all the reasonable, testable possibilities you can think of. An instructor enjoys reading these sorts of analyses because they indicate that students are thinking about what they've done. Go ahead; make an instructor happy.

Notice that in the preceding paragraph the writer did not say, "If species *X* behaves as expected, and species *Y* behaves as it did in the present experiment, then the hypothesis will be supported." This writer remembers rule number 7 (p. 5): Never make the reader back up. Notice, too, that the writer does not write, ". . . then the hypothesis will be true" or ". . . then the hypothesis will be proven." Experiments cannot *prove* anything; they can only support or disagree with hypotheses. As scientists, our interpretations of phenomena may make excellent sense based upon what we know at the moment, but those interpretations are not necessarily correct. New information often changes our interpretations of previously acquired data.

Analysis of Specific Examples

EXAMPLE 1

In this study, tobacco hornworm caterpillars were raised for four days on one diet, and then tested over a three-hour period to see if they preferred that food when given a choice of diets.

STUDENT PRESENTATION

```
The data indicate that the choice of food was not re-
lated to the food upon which they had been reared. These
data run counter to the hypothesis (Beck and Reese, 1976)
that hornworms are conditioned to respond to certain
specific foods. Only one set of data out of the four gave
any indication of a preference for the original diet, and
that indication was rather weak.
```

There are many possible explanations for data that
are so contrary to previous experimental results. Inex-
perience of the experimenters, combined with the fact
that three different people were recording data about
the worms, may account for part of the error. Keeping
track of many worms and attempting to interpret their actions as having chosen a food or having merely been pass-
ing by may have proven to be too much for first-time worm
watchers. The mere fact that each of three people will
interpret actions differently and will have somewhat
different methods of recording information introduces
bias into the data.

ANALYSIS

This Discussion section starts out well, with a comparison
between results expected and results obtained. The hypothesis being
discussed is clearly stated, and a supporting reference is given.
Moreover, the student recognizes that "data are" rather than "data
is." (On the other hand, the student persists in calling the animals
worms rather than caterpillars; the term *worms* usually refers to
annelids, whereas these animals are arthropods). The next paragraph,
however, betrays a total lack of confidence in the data obtained;
the results could not possibly have turned out this way unless the
researchers were incompetent, writes the student. Although in-
experience can certainly contribute to suspicious results, are there
no other possible explanations? Does it really take years of training
to determine whether a caterpillar is eating food *A* or food *B*?

Compare this report with the one in the next example. This
Discussion section deals with the same experiment. In fact, the
two students were laboratory partners.

EXAMPLE 2

Contrary to expectation, the results suggest that
caterpillars show no preference in the diet they touched

first and the diet they spent the most time feeding on. This unexpected finding may be due to the fact that the caterpillars were not reared on the original diets for a long enough period of time to acquire a lasting prefer‐ence. They ate only the diets they were reared on for four days, whereas the laboratory handout had suggested a pre‐feeding period of 5–10 days (Chew and Pechenik, 1984). This possibility may be tested by performing the same experiment but varying the amount of time that the caterpillars are reared on the original diets. Such an experiment would determine whether there is a critical time that caterpillars should be reared on a particular diet before they will show a preference for that diet. Another possible explanation for the results obtained may be that the caterpillars used were very young, weigh‐ing only 3–6 mg. Finally, this experiment lasted only 3 hours. Perhaps different results would have been ob‐tained had the organisms been given more time to adjust to the test conditions. It would be interesting to run the identical experiment for a longer period of time, such as 10–12 hours.

The author of this report produced a paper that clearly indicates thought. Which report do you suppose received the higher grade?

EXAMPLE 3

In this experiment, several hundred milliliters (ml) of filtered pond water was inoculated with a small population of the ciliated protozoan *Paramecium multimicronucleatum* and then distributed among three small flasks. Over the next five days, changes in the numbers of individuals per ml of water in each flask were monitored.

STUDENT PRESENTATION

The large variation observed between the groups of three replicate populations suggests that the experimental technique was imperfect. The sampling error was high because it was difficult to be precise in counting the numbers of individuals. Some animals may have been missed while others were counted repeatedly. More accurate data may be obtained if the number of samples taken is increased, especially at the higher population densities. In addition, more than three replicate populations of each treatment could be established. Finally, extremely precise microscopes and pipets could be used by experienced operators in order to reduce sampling error.

ANALYSIS

This writer, like the writer of Example 1, starts out by assuming that the experiment was a failure, and spends the rest of the report making excuses for this failure. The quality of the microscopes was certainly adequate to recognize moving objects, and *P. multimicronucleatum* was the only moving organism in the water; the author is grasping at straws. If the author had more confidence in his or her abilities, the paper might have been far different. Isn't there some chance that the experiment was performed correctly? Lacking confidence in the data, the student looked no further even though he or she actually had access to data that would have allowed several of the stated hypotheses to be assessed. On each day, for example, several sets of samples were taken from each flask, and each set gave similar estimates for numbers of organisms per ml; this consistency of results suggests that the variation in population density from flask to flask was not due to experimenter incompetence. In addition, although the student stated correctly that larger sample sizes would have been helpful, he or she should have supported that statement with additional analysis of the data. Fifteen drops were sampled from each flask for each set of samples. The student could have calculated the mean number of individuals in the first 3 drops, the first 6 drops, the first 9 drops, the first 12 drops, and then the full 15 drops, to see how estimates of population size changed as the sample size was increased. If the student had done this, he or she would probably have found that larger sample sizes are especially important when population size is least dense. (Why might this be so?)

EXAMPLE 4

In this last study, a group of students went seining for fish in a local pond. Every fish was then identified to species, so that the number of fish of each species could be determined. It turned out that 91% of the fish in the sample of 73 individuals belonged to a single species. The remaining fish were distributed among only two additional species.

STUDENT PRESENTATION

I find the small number of species represented in our sample surprising, since the pond is fed by several streams that might be expected to introduce a variety of different species into it, assuming that the streams are not polluted. The lab manual states that 12 fish species have been found in the adjacent streams. It appears that the conditions in the pond at the time of our sampling were especially suitable for one species in particular of all those that most likely have access to it. Perhaps the physical nature of the pond is such that the number of niches is small, in which case competition would become very keen; only one species can occupy a given niche at any one time (Smith, 1958). The reproductive pattern of the fishes might also contribute to the observed results. Possibly <u>Lepomis</u> <u>macrochirus</u>, the dominant spe-

cies, lays more eggs than the others, or perhaps the juveniles show better survival.

Another possible explanation for our findings might
lie in the fact that we sampled only the perimeter of the
pond, since our seining was limited to a depth of water
not exceeding the heights of the seiners. The species
distribution could be very different in the middle of the
lake at a greater depth.

ANALYSIS

I have not reproduced the entire Discussion section of the
student's paper, but even this excerpt demonstrates that a little
thinking goes a long way. Note that the student did not require
much specialized knowledge to write this Discussion section, only
a bit of confidence in the data. Another student might well have
written,

Most likely, the fish were incorrectly identified;
more species were probably present than could be recog-
nized by our inexperienced team. It is also possible that
the net had a large tear, which let most of the species
escape. I didn't notice this rip in the fabric, but my
glasses were probably dirty, and then again, I'm not very
observant.

WRITING THE INTRODUCTION SECTION

The Introduction section establishes the framework for the
entire report. In this section you briefly present background information that leads to a clear statement of the specific issue or
issues that will be addressed in the remainder of the report; by
the time you have completed writing the Materials and Methods,
Results, and Discussion sections of your laboratory report, you

should be in a good position to know what these issues are. In one or two paragraphs, then, you must present an argument explaining why the study was undertaken. More to the point, perhaps, the Introduction provides you with your first opportunity to convince your instructor that you understand why you have been asked to do the exercise.

Every topic that follows this section should be anticipated clearly in the Introduction. Conversely, the Introduction should contain only information that is directly relevant to the rest of the report.

Stating the Question

Even though the statement of questions posed, or issues addressed, generally concludes the Introduction section of a report, it is helpful in writing this section of the paper to deal with this issue first. What *was* the point of this study?

Write the following words: "In this study" or "In this experiment." Then complete the sentence as specifically as possible. Three examples follow.

In this study, the oxygen consumption of mice and rats was measured in order to investigate the relationships between metabolic rate, body weight, and body surface area.

In this study, we collected fish from two local ponds and classified each fish into its proper taxonomic category.

In this experiment, we asked the following question: Do the larvae of <u>Manduca</u> <u>sexta</u> prefer the diet upon which they have been reared when offered a choice of diets?

Note that each statement is phrased in the past tense, since the students are describing studies that have now been completed.

The strong points of these statements are best revealed by examining a few unsatisfactory ways to complete sentences dealing with the same material:

```
In this study, we measured the metabolic rate of rats and
mice.
```

```
In this study, we made a variety of measurements on fish.
```

```
In this experiment, the feeding habits of Manduca sexta
larvae were studied.
```

Each of these unsatisfactory statements is vague; the reader will assume, perhaps correctly, that you are as much in the dark about what you've done as your writing implies. Be specific. Here, in one sentence, you must come fully to grips with your experiment or study. There *was* some point to the time that you were asked to spend in the laboratory; find it.

AN ASIDE: STUDIES VERSUS EXPERIMENTS

An experiment always involves manipulating something, such as an organism, an enzyme, or the environment, in a way that will permit specific hypotheses to be tested. Containers of protozoans in pond water could be distributed among three different temperatures, for example, to test the influence of temperature on the reproductive rate of the particular species under study. As another example of an experiment, the ability of salivary amylase to function over a range of pH's might be examined to test the hypothesis that the activity of this enzyme is pH sensitive. In the field, a population of marine snails from one location might be transplanted to another location and the subsequent survival and growth of the transplanted population studied, to test the hypothesis that conditions in the new location are less hospitable for that species than in the location from which the original population was obtained. As a control, of course, the survival and growth of animals not transplanted would also have to be monitored over the same time period. Note that an experiment may be conducted in the laboratory or in the field.

It is permissible to refer to experiments as "studies," but not all studies are "experiments." In contrast to the above experiments, some exercises require you to collect, observe, enumerate, or describe. You should avoid referring to such studies as "experiments"; where there are no manipulations, there are no experiments. You might, for example, collect insects from light fixtures located at several different locations within the Biology building and identify these insects to species, enabling you to examine the distribution of insects within the building. Or you might be asked to provide a detailed description of the feeding activities of an insect. Or you might spend an afternoon documenting the depth to which light penetrates into various areas of a lake, and then correlate that information with data on the distribution of aquatic plants in the different areas. In each case, you should refer to your work as a study, not as an experiment. For example:

```
In this study, insects were collected from all light fix-
tures on floors 1, 3, and 5 of the Dana building, and the
distribution of species among the different locations
was determined.
```

Providing the Background

Having posed, in a single sentence, the question or issue that was addressed, it is relatively easy to fill in the background needed to understand why the question was asked. A few general rules should be kept in mind:

1. Back all statements of fact with a reference to your textbook, laboratory manual, outside reading, or lecture notes. Unless you are told otherwise by your instructor, do not use footnotes. Rather, refer to your reference within the text, giving the author of the source and the year of publication as in the following example:

> Many marine gastropods enclose their fertilized eggs
> within structurally complex encapsulating structures
> (Hunt, 1966; Tamarin and Carriker, 1968).

Note that the period concluding the sentence comes after the closing parenthesis.

2. Define specialized terminology. Most likely, your instructor already knows the meaning of the terms you will use, but by defining them in your own words in your report you can convince the instructor that you, too, know what these words mean. Write to illuminate, not to confuse. As always, if you write with your future self in mind as the audience, you will usually come out on top; write an introduction you will be able to understand five years from now. The following two sentences obey this, and the preceding, rule.

> A number of caterpillar species are known to exhibit in-
> duction of preference, a phenomenon in which an organism
> develops a preference for the particular flavor on which
> it has been reared (Jones and Smith, 1983).

> Molluscs are common inhabitants of the intertidal zone,
> that region of the ocean lying between the high and low
> tide marks (Jones and Smith, 1983).

3. Never set out to prove, verify, or demonstrate the truth of something. Rather, set out to test, document, or describe. In Biology (and science in general), truth is elusive; it is important to keep an open mind when you begin a study and when you write up the results of that study. It is not uncommon to repeat someone else's experiment or observations and obtain a different result or description. Responses will differ with species, time of year, and other, often subtle, changes in the conditions under which the study is conducted. To show that you had an open

mind when you undertook your study, you would want to revise the following sentences before submitting them for scrutiny by the instructor:

> In this experiment we attempted to demonstrate induction of preference in larvae of <u>Manduca</u> <u>sexta</u>.

> This study was undertaken to verify the description of feeding behavior given for <u>Manduca</u> <u>sexta</u> by Jones (1903).

> This experiment was designed to show that pepsin, an enzyme promoting protein degradation in the vertebrate stomach, functions best at a pH of 2, as commonly reported (Jones and Smith, 1983).

The first example might be modified to read,

> In this experiment, we tested the hypothesis that young caterpillars of <u>Manduca</u> <u>sexta</u> demonstrate the phenomenon of induction of preference.

How would you modify the other two examples to show that you are approaching the studies without prejudice?

4. Be brief. Include only the information that directly prepares the reader for the statement of intent, which will appear at the end of the Introduction section as discussed above. If, for example, your study was undertaken to determine which wavelengths of light are most effective in promoting photosynthesis, there is no need to describe the detailed biochemical reactions that characterize photosynthesis. As another example, consider these few sentences taken from a report describing an induction of preference study.

Caterpillars were reared on one diet for five days and tested later to see if they chose that food over foods that the caterpillars had never before experienced.

In this experiment, we explored the possibility that larvae of <u>Manduca</u> <u>sexta</u> could be induced to prefer a particular diet when later offered a choice of diets. The results of this experiment are significant because induction of preference is apparently linked to (1) the release of electrophysiological signals by sensory cells in the animal's mouth and (2) the release of particular enzymes, produced during the period of induction, that facilitate the digestion and metabolism of secondary plant compounds (laboratory handout).

The entire last sentence does not belong in the Introduction. The work referred to above was a simple behavioral study; students did not make electrophysiological recordings, nor did they isolate and characterize any enzymes. Although a consideration of these two topics might profitably be incorporated into a discussion of the results obtained, these issues should be excluded from the Introduction because they provide no rationale for this particular study; they do not explain why the study was undertaken. Include in your Introduction section only information that directly prepares the reader for the final statement of intent. You might, on a separate piece of paper, jot down other ideas that occur to you for possible use in revising your Discussion section, but if they don't make a contribution here, don't let them intrude on your Introduction. Be firm. Stay focused.

5. Write an introduction for the study that you ended up doing. Sometimes it is necessary to modify a study for a particular set of conditions, so that the observations actually made no longer

relate to the questions originally posed in your laboratory handout or laboratory manual. For example, the pH meter might not have been working on the day of your laboratory experience and your instructor modified the experiment accordingly; perhaps the experiment you actually performed dealt with the influence of temperature, rather than pH, on enzymatic reaction rates. In such an instance, you would make no mention of pH in your Introduction section, since the work you ended up doing dealt only with the effects of temperature.

The following paragraphs satisfy all the requirements of a valid introduction. This introduction is brief but complete . . . and effective.

It is well known that plants are capable of using sunlight as an energy source for carbon fixation (Jones and Smith, 1983). However, all wavelengths of light need not be equally effective in promoting such photosynthesis. Indeed, the green coloration of most leaves suggests that wavelengths of approximately 550 nm are reflected rather than absorbed, so that this wavelength would not be expected to produce much carbon fixation by green plants.

During photosynthesis, oxygen is liberated in proportion to the rate at which carbon dioxide is fixed (Jones and Smith, 1983). Thus relative rates of photosynthesis can be determined either by monitoring rates of oxygen production or by monitoring rates of carbon dioxide uptake. In this experiment, we monitored rates of oxygen production under different light conditions, to test the hypothesis that wavelengths differ in their ability to promote carbon fixation in <u>Elodea</u> <u>canadensis</u>.

DECIDING UPON A TITLE

A good title summarizes, as specifically as possible, what lies within the Introduction and Results sections of the report. Your instructor is a captive audience. In the real world of publications, however, your article will vie for attention with articles written by many other people; the busy potential reader of your paper will often glance at the title of your report and promptly decide whether to stay or move on. The more revealing your title is, the more easily your potential audience can assess the relevance of your paper to their interests. A paper that delivers something other than what is promised by the title can lose you considerable good will when read by the wrong audience and may be overlooked by the audience for which the paper was intended. Here is a list of mediocre titles, each followed by one or two more revealing counterparts:

No: Metabolic Rate Determinations
Yes: Exploring the Relationship Between Body Size and Oxygen Consumption in Mice

No: Plankton Sampling in Small Pond
Yes: Species Composition of the Spring Zooplankton of Small Pond, MA

No: Measuring the Feeding Behavior of Caterpillars
     Eating Habits of *Manduca sexta*
     Food Preferences of *Manduca sexta* Larvae
Yes: Measurements of Feeding Preferences in Tobacco Hornworm Larvae (*Manduca sexta*) Reared on Three Different Diets
     Can Larvae of *Manduca sexta* Be Induced to Prefer a Particular Diet?
     Studies on the Feeding Rates of *Manduca sexta* Larvae Given a Choice of Familiar and Novel Diets

No: Protozoan Behavior Responses
Yes: Studies on the Response of *Paramecium aurelia* to Shifts in Light and Temperature

The original titles are too vague to be compelling. Why go out of your way to give potentially interested readers an excuse to ignore your paper? Of more immediate concern in writing up laboratory reports rather than journal articles is this suggestion: Why not use a title that demonstrates to your instructor that you have understood the point of the exercise? Win your reader's confidence right at the start of your report. (By the way, the title should appear on a separate page, along with your name and the date that your report is submitted.)

WRITING AN ABSTRACT

The abstract, if requested by your instructor, is placed at the beginning of your report, immediately following the title page. Yet it should be the last thing that you write, since it must completely summarize the essence of your report: why the experiment was undertaken; what problem was addressed; how the problem was approached; what major results were found; what major conclusions were drawn. You should confine your abstract to a single paragraph, as in the example below. Abstracts are typically written in the passive voice.

This study was undertaken to determine the wavelengths of light that are most effective in promoting photosynthesis in the aquatic plant <u>Elodea</u> <u>canadensis</u>, since some wavelengths are generally more effective than others. Rate of photosynthesis was determined at 25°C, using wavelengths of 400, 450, 500, 550, 600, 650, and 700 nm and measuring the rate of oxygen production for 1 h periods at each wavelength. Oxygen production was estimated from the rate of bubble production by the submerged plant. We tested four plants at each wavelength. The rate of oxygen production at 450 nm (approximately 2.5 ml O_2/mg wet weight of plant/h) was nearly 1.5 × greater than that at any other wavelength tested, suggesting that the light of this wavelength (blue) is most readily

absorbed by the chlorophyll pigments. In contrast, light
of 550 nm (green) produced no detectable photosynthesis,
suggesting that light of this wavelength is reflected
rather than absorbed by the chlorophyll.

PREPARING THE LITERATURE CITED SECTION

In this final section of your paper, you present the complete
citations for all the factual material you refer to in the text of your
report. This presentation enables the interested reader, including,
perhaps, you at a later date, to obtain quickly the sources you
have used in preparing your report. It provides a convenient way
for the reader to obtain additional information about a particular
topic, and it also provides the reader with a means of verifying
what you have written as fact. It occasionally happens that a reference
is used incorrectly; your interpretation or recollection of what was
said in a textbook, lecture, or journal article may be wrong. By
giving the source of your information, the reader can more easily
recognize such errors. If the reader is your instructor, this list of
references may provide an opportunity for him or her to correct
any misconceptions you may have acquired. If you fail to provide
the source of your information, your instructor will have more
difficulty in determining where you went wrong. Proper referencing
is even more crucial in scientific publications. Misstatements of
fact are readily propagated in the literature by others; the Literature
Cited section of a report provides the reader with the ability to
verify all factual statements made, and the careful scientist consults
the listed references before accepting statements made by other
authors.

Listing the References

References are listed in alphabetical order according to the
last name of the first author of each publication. If you cite several
papers written by the same author, cite them chronologically. Each
citation should include the names of all authors, the year of pub-

lication, and the full title of the paper, article, or book. In addition, when citing books you must report the publisher, the place of publication, and the pages referred to. When citing journal articles, you must include the name of the journal, the volume number of the journal, and the page numbers of the article consulted.

There is unfortunately no single acceptable format for preparing this section of a report; formats differ from journal to journal. A few rules, however, do apply to most journals:

> Spell out only the last names of authors; initials are used for first and middle names.
>
> Latin names, including species names, are underlined to indicate italics.
>
> Titles of journal articles are not enclosed within quotation marks.
>
> Journal names are usually abbreviated. In particular, the word *Journal* is abbreviated as *J.*, and words ending in *-ology* are usually abbreviated as *-ol.* The *Journal of Zoology* thus becomes *J. Zool.* Do not abbreviate the names of journals whose titles are single words (for example, *Science* or *Nature*). Acceptable abbreviations for the titles of journals can usually be found within the journals themselves.

The most important rule in preparing the Literature Cited section is to provide all the information required and to be consistent in the manner in which you present it. When preparing a paper for publication, you should religiously follow the format used by the journal to which your entry will be submitted.

The following examples should be helpful in preparing the Literature Cited section of your report. Your instructor may wish to specify a different format for preparation of this section, so check on this if you are uncertain.

Literature Cited

Bayne, C. J. 1968. Histochemical studies on the egg capsules of eight gastropod molluscs. Proc. Malacol. Soc. London 38: 199–212.

Eyster, L. S. 1985. Origin, morphology and fate of the nonmineralized shell of Coryphella salmonacea, an opisthobranch gastropod. Marine Biol. 85: 67–76.

Eyster, L. S., and M. P. Morse. 1984. Early shell formation during molluscan embryogenesis, with new studies on the surf clam, Spisula solidissima. Amer. Zool. 24: 871–882.

Hunt, S. 1966. Carbohydrate and amino acid composition of the egg capsules of the whelk Buccinum undatum L. Nature, London 210: 436–437.

Hunt, S. 1971. Comparison of three extracellular structural proteins in the gastropod mollusc Buccinum undatum L. Comp. Biochem. Physiol. 40B: 37–46.

Lucas, M. I., G. Walker, D. L. Holland and D. J. Crisp. 1979. An energy budget for the free-swimming and metamorphosing larvae of Balanus balanoides (Crustacea: Cirripedia). Marine Biol. 55: 221–229.

Pitelka, D. R., and F. M. Child. 1964. Review of ciliary structure and function. In: Biochemistry and Physiology of Protozoa, Vol. 3 (S. H. Hutner, editor), Academic Press, New York, pp. 131–198.

Purves, W. K., and G. H. Orians. 1983. Life: The Science of Biology. Sinauer Associates, Inc., Boston, pp. 897–899.

A BRIEF DISCOURSE ON STATISTICAL ANALYSIS

This brief section is no alternative to a one-semester course in biostatistics; here I will merely explain why statistics are used in Biology, what is meant by the term *statistical analysis,* how the results of the analyses should be incorporated into the laboratory report, and how to talk about your data if you don't analyze them statistically.

What you need to know about tomatoes, coins, and random events. Variability is a fact of biological life: student performance on any particular examination varies among individuals; the growth rate of tomato plants varies among seedlings; the rate at which caterpillars feed on a given diet varies among individuals; the respiration rate of mice held under a given set of environmental conditions varies among individuals; the number of snails occupying a square meter of substrate varies from place to place; the amount of time a lion spends feeding varies from day to day and from lion to lion. Some of the variability we inevitably see in our data reflects unavoidable imprecision in the making of measurements. If you measure the length of a single bone 25 times to the nearest mm (millimeter), for example, you will probably not end up with 25 identical measurements. But most of the variability you record in a study reflects real biological differences among the individuals in the sample population. Variability, whether it be in the responses you measure in an experiment or in the distribution of individuals in the field, is no cause for embarrassment or dismay, but it cannot be ignored in presenting your results or in interpreting them.

Suppose we plant two groups of 30 tomato seeds on Day 0 of an experiment, and the individuals in group A receive distilled water while those in group B receive water plus a nutrient supplement. Both groups of seedlings are held at the same temperature, are given the same volume of water daily, and receive 12 hours of light and 12 hours of darkness (12L:12D) each day for 10 days. Twenty-six of the seeds sprout under the group A treatment and 23 of the seeds sprout under the group B treatment. At the end of 10 days, the height of each seedling is measured to the nearest 0.1 cm, and the data are recorded on the data sheet as shown in Figure 13. Note that the units (cm; sample size) are clearly indicated on the data sheet, as is the nature of the measurements being recorded (height after 10 days). The number of samples taken, or of measurements made, is always represented by the symbol N.

The question now is this: did the mineral supplement make a difference in the height of seedlings by day 10 after planting?

If all the group A individuals had been 2.0 cm tall and all the group B individuals had been 2.4 cm tall, we would readily

Group A seedlings: water only
(height, in cm, after 10 days)

2.1 cm, 2.1, 2.0, 2.8, 2.7, 2.4, 2.3, 2.6, 2.6, 2.5, 2.0, 2.1, 2.8, 2.0, 1.9, 2.8, 2.0, 2.2, 2.6, 1.8, 2.0, 2.2, 2.5, 2.4, 2.3, 2.1)

Average = 2.3 cm; N = 26 measurements

Group B seedlings: water plus nutrients
(height, in cm, after 10 days)

2.6 cm, 2.1, 2.0, 2.4, 2.8, 2.6, 2.6, 2.2, 2.7, 2.4, 2.4, 2.3, 2.2, 2.4, 2.6, 2.4, 2.2, 2.4, 2.8, 2.6, 2.5, 2.6, 2.4

Average = 2.4 cm; N = 23 measurements

Figure 13.   Data sheet with measurements of samples taken.

conclude that growth rates were increased by the addition of nutrients to the water. If each group *A* individual had been 2.3 cm tall and each group *B* individual had been 2.4 cm tall, we might again suggest that the nutrient supplement improved the growth rates of the seedlings. In the present case, however, there is considerable variability in the heights of the seedlings in each of the two treatments, and the difference in the average height of the two populations is not large with respect to the amount of variation found within each treatment. The heights of group *A* seedlings differ by as much as 1.0 cm (2.8 cm − 1.8 cm) and the heights of group *B* seedlings differ by as much as 0.8 cm (2.8 cm − 2.0 cm), whereas the average height difference between the two groups of seedlings is only 0.1 cm (2.4 cm − 2.3 cm).

The average height of the seedlings in the two populations is certainly different, but does that difference of 0.1 cm in average height reflect a real, biological effect of the nutrient supplement, or have we simply not planted enough seeds to be able to see past the variability inherent in individual growth rates? If we had planted only one seed in each group, the two seedlings might have both ended up at 2.6 cm; some seedlings reached this height in both treatment groups, as seen in the above listing of data. On the other hand, the one seed planted in group *A* might have been the one that grew to 2.8 cm, and the one seed planted in group *B* might have been one of the seeds that grew only to 2.2 cm. Or it might have turned out the other way around, with the tallest seedling appearing in group *B*. Clearly, a sample size of one individual in each treatment would have been inadequate to conclusively evaluate our hypothesis. Perhaps 25 seeds per sample is also inadequate. If we had planted 1,000 seeds, or 10,000 seeds in each group, the differences between the two treatments might have been even less than 0.1 cm . . . or the differences might have been substantially greater than 0.1 cm. If only we had planted more seeds, we might have more confidence in our results. If only we had measured 100,000 individuals, or one million individuals, or . . . .

But wishful thinking has little place in Biology; we have only the data before us, and they must be considered as they stand. Is the difference between an average height of 2.4 cm for the group *A* seedlings and 2.3 cm for the group *B* seedlings a real difference? That is, is the difference statistically significant? Or have we simply conducted too little sampling to see through the variability in individual results?

As another example, suppose we have mated red-eyed fruit flies with white-eyed fruit flies and, from knowledge of the parentage of these two groups of flies, we expect the offspring to have red or white eyes in the ratio of 2:1. Suppose we actually count 527 red-eyed offspring and 246 white-eyed offspring, so that slightly more than twice as many of the offspring have red eyes. Do we conclude that our expectations have been met, or that they have not been met? Is a ratio of 2.1:1 close enough to our expected ratio of 2:1? Is the result (527 red-eyed flies + 246 white-eyed flies) statistically equivalent to the expected ratio?

Biologists use statistical tests to determine the significance of differences between sampled populations, or differences between results expected and those obtained. To begin, we must precisely define a specific issue (hypothesis) to be tested. The hypothesis to be tested is called the null hypothesis, H_0. The null hypothesis always assumes that nothing unusual has happened in the experiment or study; that is, it assumes that the treatment (addition of nutrients, for example) has no effect, or that there are no differences between the results we observed and the results we expected to observe. Examples of typical null hypotheses are:

> H_0: the seedlings in groups A and B do not differ in height (or, the addition of nutrients does not alter growth rates of the seedlings).

> H_0: the eye color of offspring does not differ from the expected ratio of 2:1.

> H_0: caterpillars do not show a preference for the diet on which they have been reared.

> H_0: average wing lengths do not differ among populations of house flies.

> H_0: juniors did not do better than sophomores on the midterm examination.

It may seem surprising that the hypothesis to be tested is the one that anticipates no unusual effects; why bother doing the study if we begin by assuming that our treatment will be ineffective, or that there will be no differences in eye color, or that wing lengths will not differ from population to population? For one thing, the null hypothesis is chosen for testing because scientists must be cautious in drawing conclusions. Hypotheses can never be proven; they can only be discredited or supported, and the strongest statistical tests are those that discredit null hypotheses. The cautious approach in testing the effect of a new drug is therefore to assume that it will not cure the targeted ailment. The cautious approach in testing the effects of different diets on the growth rate or survival rate of a test organism is to assume that all diets will produce equivalent growth or survival — that is, that one diet is not superior to the others tested. The cautious approach in testing the effects of a pollutant is to assume that the substance is not harmful. Only if we can discredit the null hypothesis (the hypothesis of no effect) can we tentatively embrace an alternative hypothesis — for example, that a particular drug is effective, or that wing lengths do differ among populations, or that a pollutant is harmful.

Once we have established our null hypothesis and collected the data for our study, statistical analysis of the data can begin. A large number of statistical tests have been developed, including the familiar Chi-Square test and the Student's *t*-test. The test that should be used to examine any particular set of data will depend on the type and amount of data collected and the nature of the null hypothesis being addressed. If you are asked to conduct a statistical analysis of your data, your laboratory instructor will undoubtedly specify the test for you. Once the appropriate test is chosen, the data are maneuvered through one or more standard, prescribed formulas to calculate the desired test statistic. This test statistic may be a Chi-Square value, a *t*-value, or any of a variety of other values associated with different tests; in all cases, the calculated test value will be a single number, such as 0.93 or 129.8. A calculated value close to zero suggests that the data from the experiment are consistent with the null hypothesis (little deviation from the outcome expected if the null hypothesis is true.) A value very different from zero indicates that the null hypothesis may be wrong, since the data obtained are very different from those expected.

Returning to our seedling experiment, we wish to determine if the addition of certain nutrients alters seedling growth rates (the null hypothesis states that the nutrients have no effect). The appropriate test for this hypothesis is the *t*-test. Applying the formula provided in statistics books, the value of the *t*-statistic calculated for the data obtained in our tomato seedling experiment turns out to be −1.89. This particular value has some probability of turning up if the null hypothesis is true. Here the argument gets a bit tricky. If we repeated the experiment exactly as before, using another set of 60 seeds, we would most likely obtain a somewhat different result, and the *t*-statistic would have a different value even though the null hypothesis might still be true. If we did five identical experiments, we would probably calculate five different *t*-values from the data. In other words, a statistic may take on a broad range of values even if the null hypothesis is correct, and each of these values has some probability of turning up in any single experiment. But some values are more likely to turn up than others.

Suppose the null hypothesis, stating that the addition of nutrients does not alter the growth of tomato seedlings over the first 10 days of observation, is actually correct. If we ran our experiment (with 30 seeds planted in each of the two treatment groups) 100 times, we might actually find no measurable difference between the average heights of the seedlings in some of the experiments, so that our calculated *t*-values for these data would be zero. In most of the experiments, we would probably record small differences between the average sizes of seedlings in the two populations (and, for each of these experiments, calculate a *t*-value close to zero), and in a few experiments, purely by random chance, we would probably record large differences (and calculate *t*-values very different from zero, either much larger or much smaller). All these results are possible if we do enough experiments even though the null hypothesis is correct, simply because the growth of seedlings is variable even under a single set of experimental conditions. The oddball result may not come up very often, but there is always some probability that it will pop up in our experiment.

The important point here is that the outcome of an experiment or study can vary quite a lot, whether or not the null hypothesis is actually correct. A non-biological example may help to clarify this point. In coin tossing, a fair coin should, on average, produce an equal number of heads and tails. Yet experience tells us that 10 tosses in a row will often produce slightly more of one result than the other. Every now and then, we will actually end up tossing 10 heads in a row, or 10 tails in a row, even though the coin is perfectly legitimate; neither of these results will occur very often, but each will occur eventually if we repeat the experiment enough times.

Yes, the fact of the matter is that there is considerable morphological, physiological, and behavioral variability in the real world, and that the only way to know, with certainty, that our one experiment is a true reflection of that world is to measure or count every individual in the population under consideration (for example, plant every tomato seed in the world, and measure every seedling after 10 days) or conduct an infinite number of experiments. This is not a practical solution to the problem. The next best alternative is to use statistical analysis. Statistics cannot tell us whether we have revealed THE TRUTH, but they can indicate just how convincing or just how off-the-wall our results are.

The numerical value of any calculated test statistic has some probability of turning up when the null hypothesis is true. Statisticians tell us, for example, that values of t are distributed as in Figure 14, and that values of Chi-Square (χ^2) are distributed as in Figure 15. If the null hypothesis is correct, values of each statistic will usually fall within a certain range, as indicated; these values will have the greatest probability of turning up in any individual experiment. If the t-value calculated for our experiment falls within the range indicated as "very common," we are probably safe in accepting the null hypothesis; at least we have no reason to disbelieve it. However, even if the null hypothesis is correct, very unusual values of t or of χ^2 will occasionally occur. We are, after all, randomly picking only a few seeds to plant, out of a bag that may contain many thousands of seeds; it could be just our luck to have picked only those seeds that are most unlike the average seed.

If we calculate a very unusual (very high or very low) value for t using the data from our experiment, how can we decide to reject the null hypothesis when we know there is still some small chance that H_0 is correct and that we have simply witnessed a very rare event? Well, we must admit that we are not omniscient and be willing to take a certain amount of risk in drawing conclusions from our data; the amount of risk taken can be specified. Typically, researchers assume that if their very unusual (that is, rarely obtained) value of t (or of some other statistic) would turn up fewer than

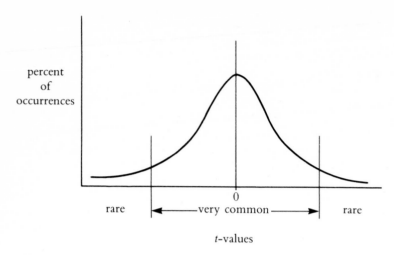

t-values

Figure 14.   The distribution of *t*-values expected when the null hypothesis (H_o) is true. A wide range of values may occur, but some values will occur more commonly than others. Obtaining a common value for *t* causes us to accept H_o. Obtaining a rare value for *t* causes us to doubt the validity of H_o.

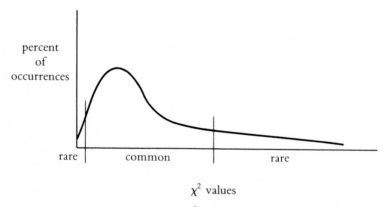

χ^2 values

Figure 15.   The distribution of χ^2 values expected when H_o is true. A wide range of values may occur, but some values will occur more often than others. The rarer the value obtained, the less confidence we can have in the validity of H_o.

5 times in 100 repetitions of the same experiment when the null hypothesis were true, then this oddball value of t is a strong argument against the null hypothesis's being correct; H_0 is then tentatively rejected. That is, the large calculated value of t would be so rarely encountered if the null hypothesis were true that the null hypothesis is *probably* wrong; but there is still the 5 in 100 chance that the null hypothesis is correct and that the researchers, through random chance, happened upon atypical results in their experiment. Tossing 10 heads in a row using a fair coin won't happen very often, but it *will* happen. Tossing 100 heads in a row is an even rarer event, but it could happen. If you conducted only one tossing experiment of 100 flips and tossed only heads, you could tentatively reject the null hypothesis that the coin is fair. But you wouldn't know with certainty that you were correct. Would you bet your car, your savings account, your little finger, or your stereo that you would get all heads if you did another round of 100 tosses? Only if the coin has two heads.

It is, I hope, becoming clear why experiments must be repeated many times before the results become convincing. Such is the challenge of doing Biology. Only when large values of a test statistic appear many times can we become fully confident that the null hypothesis deserves to be rejected. Only when low values appear many times can we become confident that the null hypothesis is most likely correct.

In summary, in performing a statistical test you first decide on a reasonable degree of risk (usually 1% or 5%) of incorrectly rejecting the null hypothesis, perform the study, plug the data into the appropriate formula to calculate the value of the appropriate statistic, and end up with a single number. You then look in the appropriate statistical table to see whether this number is within the range of values expected when the null hypothesis is correct. If your number lies within the expected range of values, your data support the null hypothesis. If your number lies outside the range of commonly expected values, your data do not support H_0; they support the alternative. But remember, there is always some small chance that your null hypothesis is correct, and that you are making the wrong decision by rejecting H_0. Similarly, when your number lies within the range of commonly recorded values, there is always some chance your null hypothesis is actually incorrect and that you are making the wrong decision by accepting it. For this reason, we biologists never seek to *prove* any particular hypothesis; we can only accumulate data that either favor or argue against the null hypothesis.

Incorporating Statistics into Your Laboratory Report

Even if you conduct no statistical analyses, be especially cautious when making statements about your data. Your data may support one hypothesis more than another, but they cannot prove that any hypothesis is true. In addition, if you conduct no statistical analyses, you cannot say that differences between groups of measurements are significant or not significant. *Significance* implies subjection of data to rigorous statistical testing. It is perfectly fair to write that "seedlings treated with nutrients appeared to grow at slightly faster rates than those treated with distilled water" and refer the reader to the appropriate table or figure, but you cannot say that seedlings in one treatment grew significantly faster than those in the second treatment.

If you have conducted statistical analysis of your data, the products of your heavy labor are readily and unceremoniously incorporated into the Results section of your report. Use the results of the analyses to support any major trends that you see in your data, as in the following two examples:

EXAMPLE 1

For thirty caterpillars reared on the mustard–flavored diet and subsequently given a choice of foods, the caterpillars showed a statistically significant preference for the mustard diet ($\chi^2 = 17.3$; $P < 0.05$). For thirty caterpillars reared on the quinine–flavored diet, there was no influence of previous experience on the choice of food ($\chi^2 = 0.12$; $P > 0.10$).

EXAMPLE 2

Over the first 10 days of observation, growth of seedlings receiving the nutrient supplement was not significantly faster than the growth of seedlings receiving only water (t = -1.89; P > 0.10).

In the first example, H_0 states that prior experience will not influence the subsequent choice of food by caterpillars; "P < 0.05" means that if the experiment were repeated 100 times and H_0 were true, such a high value for χ^2 would be expected to occur in fewer than 5 of those 100 studies. In other words, the probability of making the mistake of rejecting H_0 when it is, in fact, true is less than 5%. You can therefore feel reasonably safe in rejecting H_0 in favor of the alternative: that prior experience does influence subsequent food selection for caterpillars reared on the mustard diet.

Different results were obtained, however, for the caterpillars reared on the quinine-flavored diet. In example 1, sentence 2, "P > 0.10" means that if the experiment were repeated 100 times and H_0 were true, you would expect to calculate such a small value of t in at least 10 of the 100 trials. In other words, the probability of getting this t-value with H_0 true is rather high; certainly the t-value is not unusual enough for you to mistrust H_0 and run the risk of rejecting the null hypothesis when it is, in fact, true.

In the second example, the null hypothesis (H_0) states that the nutrient supplement does not influence plant growth; "P > 0.10" again means that if the experiment were repeated 100 times and H_0 were true, you would expect to calculate such a low value of t in more than 10 of the 100 trials. As before, you have obtained a value of t that would be common if H_0 were true and so have no reason to reject H_0. It is, of course, possible that H_0 is actually false and the nutrients really do promote seedling growth, and that you just happened upon an unusual set of samples that gave a misleadingly small t-value. If such is the case, repetition of the experiment should produce different results and larger t-values. But with only the data before you, you cannot reject H_0.

Note that you need say little about the statistics themselves when writing your report. You would simply state, in your Methods section, that the data were analyzed by Chi-Square or some other test, and then, in the Results section, include a few test statistics to back up your interpretations of data as indicated above; the few sentences of Examples 1 and 2 tell the complete story. Statistics are used only to back up any claims you wish to make; resist the temptation to ramble on about how the statistics were calculated, how brilliant you are to have figured out which calculations to make, or how awful it was to make them.

3
Writing Essays and Term Papers

A term paper is really just a long essay, its greater length reflecting more extensive treatment of a broader issue. Both assignments present critical evaluations of what has been read. In preparing an essay, you synthesize information, explore relationships, analyze, compare, contrast, evaluate, and organize your own arguments clearly, logically, and persuasively, gradually leading up to an assessment of your own. A good term paper or short essay is a creative work; you must interpret thoughtfully what you have read and come up with something that goes beyond what is presented in any single article or book consulted.

Essays and term papers are based mostly on readings from the primary scientific literature — that is, the original research papers published in such scientific journals as *Biological Bulletin, Developmental Biology, Ecology,* and *Journal of Comparative Biochemistry and Physiology*. Textbooks and review articles (such as those in *Scientific American*) compose the secondary literature. The secondary literature gives someone else's interpretation and evaluation of the primary literature. In preparing an essay or term paper, you will go through the same processes that the writers of textbooks and review articles go through in presenting and discussing the primary scientific literature.

WHY BOTHER?

Every time you are asked to write an essay or term paper, your instructor is committing himself or herself to many hours of reading and grading. There must be a good reason to require such assignments; generally speaking, most instructors are not masochists.

In fact, writing essays or term papers benefits you in several important ways. For one thing, you end up teaching yourself something relevant to the course you are taking. The ability to self-teach is essential for success in graduate programs and academic careers, and is a skill worth cultivating for success in almost any profession. Additionally, you gain experience in reading the primary scientific literature. Textbooks and many lectures present you with facts and interpretations. By reading the papers upon which these facts and interpretations are based, you come face to face with the sorts of data, and interpretations of data, that the so-called facts of Biology are based on, and you gain insight into the true nature of scientific inquiry. The data collected in an experiment are always real; interpretations, however, are always subject to change. Preparing thoughtful essays and term papers will help you move away from the unscientific, blind acceptance of stated facts towards the scientific, critical evaluation of data. These assignments are also superb exercises in the logical organization, effective presentation, and discussion of information, skills that can only ease your career progress in the future. How fortunate you are that your instructor cares enough about your future to give such assignments!

There is one last reason that instructors often ask their students to prepare essays. One can simply summarize the contents of a dozen papers in succession without understanding the contents of any of them. I call this the book report format, in which the writer merely presents facts uncritically: this happened, that happened; the authors suggested this; the authors found that. By writing an essay rather than a book report, you can show your instructor that you really understand what you have read, that you have really learned something rather than simply memorized or mimicked the information presented to you.

GETTING STARTED

You must first decide on a general subject of interest. Often your instructor will suggest topics that have been successfully exploited by former students. Use these suggestions as guides, but do not feel compelled to select one of these topics unless so instructed. Be sure to choose or develop a subject that interests you. It is much easier to write successfully about something of interest than about something that bores you.

All you need for getting started is a general subject, not a specific topic. Stay flexible. As you research your selected subject, you will usually find that you must narrow your focus to a particular topic because you encounter an unmanageable number of references pertinent to your original idea. You cannot, for instance, write about the entire field of primate behavior because the field has many different facets, each associated with a large and growing literature. In such a case, you will find a smaller topic, such as the social significance of primate grooming behavior, to be more appropriate; as you continue your literature search, you may even find it necessary to restrict your attention to a few primate species.

Alternatively, you may find that the topic originally selected is too narrow, and that you cannot find enough information to base a substantial paper on. You must then broaden your topic, or switch topics entirely, so that you will end up with something to discuss. Don't be afraid to discard a topic on which you can find too little information.

Choose a topic you can understand fully. You can't possibly write clearly and convincingly on something beyond your grasp. Don't set out to impress your instructor with complexity; instead, dazzle your instructor with clarity and understanding. Simple topics often make the best ones for essays.

RESEARCHING YOUR TOPIC

Begin by carefully reading the appropriate section of your textbook to get an overview of the general subject of which your topic is a part. It is usually wise to then consult one or two additional textbooks before venturing into the primary literature; a solid construction requires a firm foundation. Your instructor may have placed a number of pertinent textbooks on reserve in your college library. Alternatively, you can consult the library card file, looking for books listed under the topic you have chosen to investigate.

You have to become a little devious at times before you can convince the card file to satisfy your request for information. Suppose, for example, that you wish to find material on reptilian respiratory mechanisms. You might try, to no avail, looking under Respiration or Reptiles, but looking under Physiology or Comparative Physiology will probably pay off. Similarly, in researching the topic of annelid locomotion you might try, unprofitably, searching under Annelids, Locomotion, or Worms; looking under Invertebrate Zoology will probably turn up something useful. Using a library card file is like using the Yellow Pages of the telephone book; the phone number of the local movie house isn't found under a heading of Movies or Movie Theaters, but under Theaters, not the first place I'd look. If at first you don't succeed . . .

The references given in textbooks often provide good access to the primary literature, as do those given in review articles. Two journals in particular may contain especially useful reviews in this regard: *Scientific American* and *Biological Reviews*. It is also profitable to browse through recent issues of journals relevant to your topic; ask your instructor to name a few journals worth looking over. If you find an appropriate article in the recent literature, consult the literature citations at the end of the article for additional references worth consulting. This is an especially easy and efficient way to accumulate research material; the yield of good references is usually high for the amount of time invested.

Another way to track down pertinent recent references is by using the *Science Citation Index,* published by the Institute for Scientific Information (ISI), Incorporated, in Philadelphia. Because it is expensive, not all libraries subscribe to this service, so be sure to check with your reference librarian before getting your hopes up. Your reference librarian should also be able to assist you in using any of the services discussed in this section.

To use the *Science Citation Index* you must first have discovered at least one paper (the so-called key paper) from the primary literature pertinent to your quest. Consulting the Citation volumes of the index, you look under the name of the author who wrote this key paper. Below that author's name you will find a listing of additional references, one of which should be for the paper you have already read and found to be particularly appropriate to your topic. Beneath the listing for this reference you will find a list of all the papers that have cited your key paper during the year covered by the index volume consulted. Suppose, for example, you have obtained and read the following reference, cited at the end of a chapter in your class textbook, and have determined the paper to be of special use in developing your topic:

> Pearse, V. B. 1974. Modification of sea anemone behavior by symbiotic zooxanthellae: phototaxis. *Biological Bulletin* (Woods Hole) 147: 630–640.

Looking in the Citation volume for 1983, you will find the listing shown in Figure 16.

I have indicated the paper of interest to us by an arrow at the left of the reference:

> 74 Biol B Woods Hole 147 630.

Below this reference we find one paper listed:

> SEBENS KP ECOL MONOGR 53 405 83

This listing tells us that a paper citing the Pearse (1974) article was published by K. P. Sebens in *Ecological Monographs,* volume 53, beginning on page 405; the Sebens paper was published in 1983. Most likely, a paper that cites your key reference in its bibliography is appropriate to your topic and is therefore worth consulting.

Science Citation Index is published at two month intervals and is consolidated into a smaller number of volumes yearly.

Another useful indexing service is the *Zoological Record,* published jointly by BioSciences Information Service and the Zoological Society of London. *Zoological Record* is published once a year in several volumes, and each volume is devoted to a particular animal phylum or group of related phyla. You will, for example, find separate volumes devoted to the Mollusca, the Annelida, and the

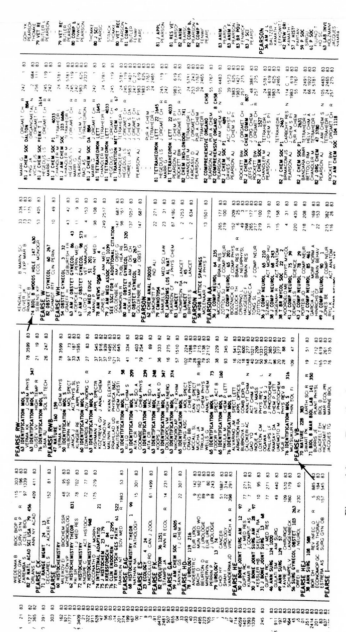

Figure 16.   Detail of a page of the Citation volume from the 1984 *Science Citation Index.*

Chordata. At the front of each volume is a section arranged by subject, as shown in the example in Figure 17 taken from the 1984 volume on the Mollusca.

Suppose we wish to find some references concerning the influences of hydrocarbon pollutants on the physiology of marine clams. As indicated by the arrow in Figure 17, the subject *Pollution* is probably a good bet, so we then turn to page 184 of the same volume. On page 184 of *Zoological Record* (see Figure 18) we find a variety of interesting references, all of which relate to our topic and most of which were published in the single year covered by the volume of *Zoological Record* we are consulting; in this case, most of the papers were published in 1982. Under the heading *Chemical Pollution,* a paper by Mix *et al.* seems particularly promising, and we are referred to reference no. 2584, as indicated by the arrow. The complete reference for this paper, including names of all authors, the title of the paper, the name of the journal, and page numbers on which the article appears, is given towards the front of the volume, as shown in Figure 19. (Note the arrow in front of the citation for the Mix article.) These citations are listed alphabetically and also numerically, so that we could have looked up the reference under "Mix" rather than by number. *Zoological Record* is a very valuable source of information. (It takes some time to put something like this together, which is why the 1984 volume covers papers published in 1982.)

Another widely used service is *Biological Abstracts,* published by BioSciences Information Service in Philadelphia. *Biological Abstracts* is the most up-to-date of the published indexing services, but it can be frustrating to use. You begin using *Biological Abstracts* in a straightforward manner, by looking up key words relevant to the topic being researched. Suppose, for example, you are looking for papers discussing the effect of histamine on blood circulation in humans. Key words, such as *histamine,* are listed in the central portion of each column, as shown in Figure 20.

At this point, *Biological Abstracts* becomes a mixed blessing. On the one hand, *Biological Abstracts* is only a few months behind in compiling the current literature, whereas *Science Citation Index* and *Zoological Record* are one to two years behind. On the other hand, an appropriate key word in *Biological Abstracts* doesn't always lead to a useful reference. To the right and left of the key word

Nova Scotia	332	Pigments	258	Reproductive system
Nucleolus	198	Piscean hosts	308	Reproductive technique
Nucleoplasm	198	Plankton	288	Republic of Ireland
Nucleotide function	246	Plankton collection	189	Research reports
Nucleus	198	Plant & vegetation habitats	295	Reservoir
Number of generations	264	Plasma	234	Respiratory adaptations
Nutrition	230	Plasma membrane	198	Respiratory function
		Pleistocene	360	Respiratory gas exchange
Obituaries	188	Pliocene	361	Respiratory gas transport
Oceanic islands	339	Poisonous animals	183	Respiratory pigments
Ohio	335	Poland	322	Respiratory quotient
Oklahoma	335	Pollutant content	244	Respiratory rate
Oligocene	363	Pollutant metabolism	250	Respiratory system
Ontario	332	Pollution	184	Respiratory water current
Ontogenesis	269	Polymorphism	277	Resting behaviour
Oogenesis	261	Polyploidy	275	Retina
Operculum	199	Pond	294	Ribosomes
Optomotor response	299	Population censuses	287	Rickettsial diseases
Ordovician	371	Population changes	285	Ritualised behaviour
Oregon	336	Population density	284	River
Organelles	198	Population density measurement	192	River Danube
Organic chemicals	253	Population dynamics	283	River Mekong
Organized study	183	Population energetics	280	River Nile
Oriental region	327	Population genetics	273	River Rhine
Orientation	302	Population inbreeding	273	River St Lawrence
Origin of taxon	275	Population mapping	192	Rock habitat
Osmotic relations	240	Population regulation	285	Rocky substrate
Osmotrophic nutrition	230	Population sampling	192	Role of biomembrane
Osphradium	219	Population sex ratio	283	Romania
Ovarian cycle	263	Population size	284	Rostrum
Ovary	263	Population structure	282	Russian SFSR

Figure 17.　Detail of a page of subject listings on the Mollusca from *Zoological Record* of 1984.

BIOLOGICAL POLLUTION

Bacteria, condition index relationship, South Carolina (marine)
Crassostrea virginica   SCOTT, G.I., ET AL (3404)
Contamination by human faecal bacteria, lagoon, France
Mytilus galloprovincialis   BALEUX, B., ET AL (243)
Sewage, effect on intertidal algal mat abundance
Significance for benthic community ecology, England
    NICHOLLS, D.J., ET AL (2759)

CHEMICAL POLLUTION

Aromatic hydrocarbons, effect on enzyme activity, Shetlands (marine)
Littorina littorea   MOORE, M.N., ET AL (2619)
Arsenic, Scotland (marine)
Littorina   SHEPHERD, R.J., ET AL (3450)
Arsenic & trace elements, seasonal variation, Oregon (marine)
Mytilus edulis   LA TOUCHE, Y.D., ET AL (2068)
Behaviour, physiological condition & oxygen consumption, relationships
Bullia   BROWN, A.C. (526)
Benzo(a)pyrene accumulation, Black Sea
Mytilus galloprovincialis   ZOBOVA, N.A. (4254)
Benzo(a)pyrene seasonal content in various tissues, Oregon (marine)
Mytilus edulis   MIX, M.C., ET AL (2584)
BHC & DDT pesticide residue determination
Meretrix   ZHANG, T., ET AL (4243)
Biomass & population density relationship, Arabian Sea
Cellana radiata   PRASAD, M.N., ET AL (3053)
Brackish habitat adaptations, Italy

ce between D. Dupuy & A.L. Montandon,
    VAN DOREN, P. (3925)
    ANDREI, G., ET AL (152)

iccount
ora   ZANINETTI, L. (4229)
ription by Aristotle
    LANDMAN, N.H. (2085)
    LANDMAN, N.H. (2085)
gium   DUCHAMPS, R. (1057)

)IFICATION

cture & population density
nipulation effects, Utah
    WILLIAMS, R.D., ET AL (4121)
nce, ecological reaction, Czechoslovakia
    MACHA, S. (2289)
causes, effects on fauna, Gulf of California
    POORMAN, L.H. (3028)
on, community changes, USA
    TAYLOR, R.W., ET AL (3770)
level increase, effect on community structure.
    SHEPARD, W.D., ET AL (3448)
effect on community structure, River St
    HAYNES, J.M., ET AL (1563)
on effect on river community, Arkansas &
    GORDON, M.E. (1406)

Figure 18.   Detail of page 184 of 1984 *Zoological Record* showing references on chemical pollution.

Mitra, H. Chandra *see* Sahni, A.

Mitra, S.C. *see* Subba Rao, N.V.

Mitsuhashi, T. *see* Endo, S.

Mitsuhashi, T. *see* Endoh, S.

Mix, M.C., Hemingway, S.J. & Schaffer, R.L. (2584)
Benzo(a)pyrene concentrations in somatic and gonad tissues of bay mussels, *Mytilus edulis.*
Bulletin envir. Contam. Toxicol. 28(1) 1982: 46-51, illustr.
[In English]

Mix, M.C. *see* la Touche, Y.D.

Mix, M.C. *see* Latouche, Y.D.

Miyamoto, K., Inaoka, T., Hayasaka, K., Kutsumi, H., Oku, Y. & Yagi, K. (2585)
Studies on the zoonoses in Hokkaido. 4. Detection of *Metagonimus yokogawai* cercariae from the fresh water snail (*Semisulcospira libertina* Gould, 1895) in Asahikawa and Sapporo Cities.
Japanese J. Parasit. 31(5) 1982: 377-384, illustr.
[In Japanese with English summary]

Miyamoto, T., Saito, K., Motoya, S., Nishikawa, N., Monma, H. & Kawamura, K. (2586)
Experimental studies on the release of the cultured seeds of abalone, *Haliotis discus hannai* Ino in Oshoro Bay, Hokkaido.
Scientific Rep. Hokkaido Fish. expl Stn No. 24 1982: 59-89,

Mohamed, A.M. & Ishak, M.M.
Comparative studies on the effects of castratio
schistosome-infection on *Bulinus truncatus.*
Journal Egypt. Soc. Parasit. 12(2) 1982: 499-512, illustr.
[In English]

Mohamed, A.M. & Ishak, M.M.
Seasonal changes in the physiological activity of land
Helicella vestalis (Gastropoda, Stylommatophora) in Egy
Cellular molec. Biol. 27(5) 1981: 557-561, illustr.
[In English with French summary]

Mohan, M.V.
Allometric relationships in the green mussel, *Perna virid*
Indian J. mar. Sci. 9(3) 1980: 224-226, illustr.
[In English]

Mohan, M.V. & Cheriyan, P.V.
Oxygen consumption of *Nausitora hedleyi* Schepman in rela
to salinity.
Indian J. mar. Sci. 11(3) 1982: 278-280, illustr.
[In English]

Mohan, M.V. & Cheriyan, P.V.
Oxygen consumption of *Nausitora hedleyi* Schepman and
furcifera von Martens in relation to oxygen tension.
Bulletin Dep. mar. Sci. Univ. Cochin 11(2) 1980: 77-88, ill
[In English]

Mohan, M.V. & Cheriyan, P.V.
Oxygen consumption of the green mussel, *Perna viridis* L
relation to body weight and declining oxygen tension.

Figure 19.   Detail of page from 1984 *Zoological Record* showing complete reference for the paper by Mix *et al.*

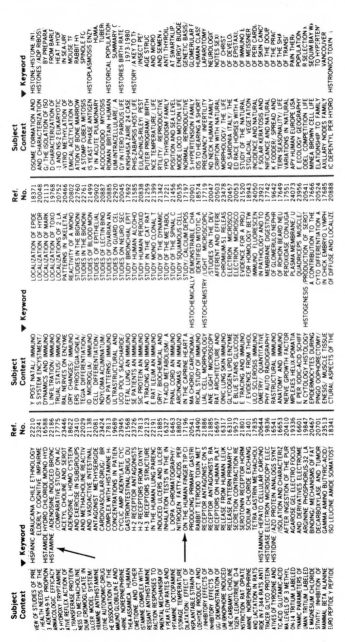

Figure 20. Detail of a page from *Biological Abstracts* listing key word *histamine* and related reference numbers.

histamine, for example, and for the thirty-two lines under this key word, you will find a few words or parts of words (drawn from each paper's title) that partially inform us of each paper's content. The only way to tell whether or not a reference is relevant to our topic is to look it up, as discussed below. Such listings provide many ambiguous leads, each of which must be investigated. One entry, indicated in Figure 20 by an arrow, does appear somewhat promising:

DILATION/EFFECT OF. . .ON THE HUMAN FINGER TIP CI.

Perhaps the *CI* stands for *circulation?* Ever hopeful, we turn to reference number 17196 as directed and are rewarded with the listing indicated with an arrow in Figure 21. Not only does *Biological Abstracts* provide us with the complete citation for the paper by Coffman, Cohen, and Rasmussen, but it gives us an informative abstract as well, summarizing the contents of the paper.

Biological Abstracts is currently published in 24 bi-monthly issues compiled in 2 volumes per year; each is about 2 inches thick — which gives you some idea of the tremendous rate at which research articles are now being published.

Lastly, I should mention the existence of computerized search services. Through these services you can enter a series of key words and subjects and have the computer search its data banks for relevant papers published within whatever time-frame you specify. In a matter of minutes, the computer will examine from ten to twenty years of source material and give you a list of all references appropriate to the information provided. Computer searches are expensive to run, however, and often require specialized training before they can be used effectively. For these reasons, few undergraduates will have access to this search tool.

Science Citation Index, Zoological Record, Biological Abstracts, and computer search systems are all good sources of references, but there is a major catch; your library will probably not subscribe to all the journals included in the literature searched by the various services. You may thus spend considerable time accumulating a long list of intriguing references, only to discover that most of them are not to be found on your campus or in any other nearby

TSIBEZOV, V. V. (Cent. Res. Lab., Fourth Main Adm., Minist.
R, Moscow, USSR.) BIOKHIMIYA 48(8): 1384–1389. 1983. [In
Russ. and Engl. summ.] **Some properties of rat heart peptidase
luliberin.**—The properties of rat heart peptidase hydrolyzing
[RH] were studied. This peptidase was a sulfhydryl metalloenzyme
f about 100,000. The maximal enzyme activity was observed at
.s of pH, Ca^{2+} (5 · 10⁻⁶ *M*) increased the enzyme activity by 50%,
indicative of an anomalous dependence of the enzyme activity of
concentration. At luliberin concentrations of 10^{-7}–10^{-6} *M* the enzyme
by Ca^{2+} was considerably reduced and returned to the initial level
eptide concentration was increased up to 10⁻⁵ *M*. Evidently, the
nder study is a regulatory enzyme whose activity depends on
ons of Ca^{2+} and of the reaction substrate, luliberin.

FARIS, BARBARA*, ROCCO FERRERA, PAUL TOSELLI,
MBU, WAYNE A. GONNERMAN and CARL FRANZBLAU.
v. Sch. Med., Dep. Biochem., 80 East Concord St., Boston, MA
CHIM BIOPHYS ACTA 797(1): 71–75. 1984. **Effect of varying
scorbate on collagen, elastin and lysyl oxidase synthesis in aortic
le cell cultures.**—In the presence of ascorbate, there is an increase
synthesis with a concomitant decrease in insoluble elastin and lysyl
tivity in cultured rabbit aortic smooth muscle cells. While the
r 0.5 μg ascorbate/ml of medium enhances collagen synthesis and
n, detectable insoluble elastin and lysyl oxidase activity remain
nchanged. At 2 μg ascorbate/ml, the integrity of the insoluble
st and lysyl oxidase activity is decreased. Evidently, by modifying
of ascorbate in the culture medium the nature of the extracellular
duced by the smooth muscle cells can be altered.

ARUMUGHAM, R. and JOHN C. ANDERSON*. (Dep. Biochem.,
, Univ. Manchester, Manchester M13 9PT.) BIOCHIM BIOPHYS

→ **17196.** COFFMAN, JAY D.*, RICHARD A. COHEN and HELEN M.
RASMUSSEN. (Univ. Hosp., 75 East Newton St., Boston, Mass. 02118,
USA.) CLIN SCI (LOND) 66(3): 343–350. 1984. **Effect of histamine on the
human fingertip circulation.**—The effect of intraarterial histamine on fingertip
blood flow (FBF) and vascular resistance (FVR) was studied in normal
subjects during reflex sympathetic vasoconstriction induced by body cooling
and vasoconstriction caused by intraarterial noradrenaline [norepinephrine]. In
a room at 20° C, FBF increased from 15.3 ± 35.5 (SD) to 28.3 ± 55.9 ml
min⁻¹ 100 ml⁻¹ of tissue and FVR decreased from 23.7 ± 17.7 to 11.9 ± 9.9
mm Hg · ml min⁻¹ 100 ml⁻¹ ($P < 0.01$) during infusions of histamine (0.5–4
μg/min) in 14 subjects. In 9 of these subjects, the disappearance half times of
local injections of $Na^{131}I$ were measured and decreased from 19.8 ± 10.9 to
12.9 ± 7.3 min during histamine infusions, indicating an increase in nutritional
flow. Arteriovenous shunt flow was also probably affected for increases in FBF
were sometimes large and FBF increased without a change in the radioisotope
half time in 2 subjects. Neither cimetidine nor pyrilamine (mepyramine)
consistently prevented the FBF responses to histamine. Administration of both
antihistamines together attenuated the response. During noradrenaline infusions
in 4 subjects, a large increase in FBF (8.9 ± 10.9 to 39.0 ± 8.2 ml, $P < 0.005$)
occurred at the smallest dose (0.5 μg/min) of histamine. Histamine can
vasodilate fingertips and increase nutritional blood flow during reflex sympathetic
vasoconstriction. This vasodilatation may be mediated via both histamine H1
and H2 receptors.

17197. NAGPAL, S. K*., B. S. NANDA, L. N. DAS and R. P.
SAIGAL*. (Dep. Anat. and Histol., Punjab Agric. Univ., Ludhiana.)
J RES PUNJAB AGRIC UNIV 20(2): 213–216. 1983[recd. 1984]. **Calcium
deposition as an age associated change in the caprine heart: A histochemical
study.**—Deposition of calcium salts in the heart and coronary arteries of goats
of 2 days to > 6 yr of age was studied histochemically. The youngest animal
showing Ca deposition was 4 mo. of age. Thereafter a progressive increase was

Figure 21. Detail of page from *Biological Abstracts* with complete citation of paper by Coffman *et al.*, and summary of the paper's
contents.

library. Consulting recent issues of available, appropriate journals may thus be the most efficient way to search for promising research topics and references.

PLAGIARISM AND NOTE-TAKING

The essay or term paper you submit for evaluation must be original work: yours. Submitting anyone else's work under your name is plagiarism and can get you expelled from college. Presenting someone else's ideas as your own is also plagiarism. Consider the following two paragraphs.

Smith (1981) suggests that this discrepancy in feeding rates may reflect differences in light levels used in the two different experiments. Jones (1984), however, found that light level did not influence the feeding rates of these animals and suggested that the rate differences reflect differences in the density at which the animals were held during the two experiments.

This discrepancy in feeding rates might reflect differences in light levels. Jones (1984), however, found that light level did not influence feeding rates. Perhaps the difference in rates reflects differences in the density at which the animals were held during the two experiments.

The first example is fine. In the second example, however, the writer takes credit for the ideas of Smith and Jones; the writer has plagiarized.

Plagiarism sometimes occurs unintentionally, through faulty note-taking. Photocopying an article or book chapter does not constitute note-taking; neither does copying a passage by hand, occasionally substituting a synonym for a word used by the source's author. Take notes using your own words; you must get away from being awed by other people's words and move towards building confidence in your own thoughts and phrasings. Note-taking involves critical evaluation; as you read, you must decide either that particular facts or ideas are relevant to your topic or that they are irrelevant. As Sylvan Barnet says in *A Short Guide to Writing About Art* (1981. Little, Brown and Company, second edition, p. 142), "You are not doing stenography; rather, you are assimilating knowledge and you are thinking, and so for the most part your source should be digested rather than engorged whole." If an idea is relevant, you should jot down a summary using your own words. Try not to write complete sentences as you take notes; this will help you avoid unintentional plagiarism later and will encourage you to see through to the essence of a statement while note-taking.

Sometimes the authors' words seem so perfect that you cannot see how they might be revised to best advantage for your paper. In this case, you may wish to copy a phrase or a sentence or two verbatim, but be sure to enclose this material in quotation marks as you write, and clearly indicate the source and page number from which the quotation derives. If you modify the original wording slightly as you take notes, you should indicate this as well, perhaps by using modified quotation marks: ^{ʕʕ} . . . . . ^{ʕʕ} . If your notes on a particular passage are in your own words, you should also indicate this as you write. I precede such notes, reflecting my own ideas or my own choice of words, with the word *Me* and a colon; my wife, who is also a biologist, uses her initials. If you take notes in this manner you will avoid the unintentional plagiarism that occurs when you later forget who is actually responsible for the wording of your notes, or who is actually responsible for the origin of an idea.

Here is an example of some notes taken using this system of notation. These notes are based on a paper published by R. A. Merz (1984): Self-generated *versus* environmentally produced feeding currents: a comparison for the sabellid polychaete *Eudistylia vancouveri*. *Biological Bulletin* (Woods Hole) 167:200–209. (See Figure 22.) In the example shown in Figure 23, the note-taker has clearly

The mechanisms by which suspension feeding animals remove particles from the surrounding fluid is a topic of current and historical interest. Fluid movement determines to a large degree the mechanical forces impinging on an organism (*e.g.*, Wainwright and Koehl, 1976; Merz, 1984), the rates of respiration and excretion (LaBarbera, 1982), and the feeding mode employed by some organisms (*e.g.*, Lewis, 1968; Warner, 1977; Tagon *et al.*, 1980; LaBarbera, 1984). Therefore, to fully and accurately understand the feeding processes and behavior of aquatic organisms, the natural flow regime of the animal in question must be taken into account (Reidl, 1971; Vogel, 1981).

Nicol (1930) describes the morphology, ciliary tracts, and feeding currents of the sabellid polychaete *Sabella pavonina*. This very detailed work is one of the most complete descriptions of the feeding mechanisms of a polychaete (Fauchald and Jumars, 1979) and has been used as a model for other studies of sabellid polychaetes (Fitzsimmons, 1965; Lewis, 1968; Bonar, 1972). It has also been incorporated into the literature as a general model for feeding in the Sabellidae (Jørgenson, 1956, 1966; Dales, 1970; Barnes, 1980). Nicol suggests that all water movement through the branchial crown of sabellids is due to ciliary activity. However, Nicol did not account for possible effects that ambient flow may have in this process. Her observations on whole worms were carried out in small closed containers of still water; finer details were ascertained by examining excised portions of branchiae.

Dales (1957) estimated the filtration rate (volume of material strained per unit fresh worm weight) of a variety of fan worms (sabellids and serpulids) in still water by measuring changes in the optical density of graphite particles and unicellular algal suspensions. He concluded that fan worms are "clearly . . . less efficient than other suspension-feeding invertebrates, both in the volume of water they are capable of straining, and in the kind of particles which can be retained" (p. 315). However, Warner (1977) suggested that sabellids are among the suspension feeders which can use ambient water movement to augment their own self-produced currents. Because the effect of the environmental regime has never been addressed in any study of sabellids, an important aspect of the feeding mechanisms and ecology of these animals has been neglected.

Many species within Sabellidae aggregate into densely packed nonclonal mounds (Hartman, 1969). Aggregations of tube-dwellers have been shown to affect the pattern of ambient currents (Eckman, 1979, 1983; Nowell and Church, 1979). The amount of suspended material in the water may be augmented by resuspension of particles due to the presence of tubes (Eckman *et al.*, 1981; Carey, 1983) or may be depleted by the biological activity of the tube dwellers (Fager, 1964; Woodin, 1978, 1981; Levin, 1982). None of these studies has examined these phenomena for epifaunal tubes on hard substrates.

This work examines three aspects of suspension feeding in the sabellid polychaete, *Eudistylia vancouveri*. First is a comparison of the water velocities produced by the cilia of the branchial crown with the velocities of ambient currents. Second, the effect of the dense hemispherical aggregates of worms on the water flow near the feeding crown is described. Third, removal of natural particles during a single passage of water across the surface of a worm cluster in the field is quantified.

Field site and animal collection

Cattle Point, San Juan Island, Washington (48°27'N, 122°57'W) was the site for all *in situ* flow measurements, worm collection, and particle sampling. This rocky point extends into the Strait of Juan De Fuca and is one of the most exposed points in the San Juan Islands.

Specimens of *E. vancouveri* were collected by carefully peeling intact tubes away from the rock substrate during low tides. The animals were held in sea water tables with continuous circulation of fresh sea water. Only whole, undamaged worms were used for flow observation and measurement.

Flow observation and measurement

To measure and observe currents produced by cilia of the branchial crown, individual worms were supported upright in their natural tubes in a $15 \times 10 \times 20$ cm clear plastic container and fresh sea water was circulated through the space between this inner container and an outer chamber to keep the inner box at ambient sea water temperature (10°–12°C).

Flow patterns produced by the branchial crown were visualized by releasing fluorescein sodium (uranine) dye (dissolved in sea water) at various locations around the worm. This dye was injected through PE-50 catheter tubing, the end of which

Figure 22.   Facsimile of article by R. A. Merz (1984) from which notes were taken (see Figure 23).

① Nicol (1930) paper = "One of the most complete descriptions polych. feeding mech." Yet, even this paper studied feeding in still water, not a natural sit. for these worms. me: see this Nicol ref, Trans. R. Soc. Edinb. 56:532.

② Sabellid polych. generally cluster → clusters of some species previously shown to affect local water circ. patterns. Effect not yet quantified for *E. vancouveri*. Common assumption that cilia on worm tentacles are completely responsible for water movement around worms may be wrong.

③ Lab/field study. In lab, used ambient seawater temps (10-12°C). me: Perhaps could get different results if did expts at different temps?

④ Revealed water circ. patterns in still water (cilia influence only) by adding small amounts of fluorescein dye at different positions near worm; monitored dye movement.

Figure 23.   Handwritten notes based on article by R. A. Merz (see Figure 22).

distinguished between his or her thoughts and the author's thoughts, and between what the author has done and what the student thinks could be done later or might have influenced the results ("could get different results at different temperatures?"). Note that the student has avoided using complete sentences, focusing instead on getting the basic points and pinning down a few words and phrases that might be useful later. The student will not have to worry about accidental plagiarism when writing a paper based on these notes. Moreover, the student is well on the way to preparing a solid essay, since the style of note-taking indicates clearly that the student has been thinking while reading.

Some people suggest taking notes on index cards, with one idea per card so that the notes can be sorted readily into categories at a later stage of the paper's development. If you prefer to take notes on full-sized paper, beginning a separate page for each new source and writing on only one side of each page will facilitate sorting later.

As you take notes, be sure to make a complete record of each source used: author(s), year of publication, volume and page numbers (if consulting a scientific journal), title of article or book, publisher, and total number of pages (if consulting a book). It is not always easy to relocate a source once returned to the library stacks; the source you forget to record completely is always the one that vanishes as soon as you realize that you need it again. Also, before you finish with a source, it is good practice to read the source through one last time to be sure that your notes accurately reflect the content of what you have read.

As another example of effective note-taking, consider some notes based on several paragraphs from Charles Darwin's *The Origin of Species,* published in 1859. Accompanying the notes is the selection from Darwin's work on which the notes were based (Figure 24). The notes (Figure 25) were being taken for an essay on the mechanism of natural selection. Another passage from Darwin appears in Figure 26, with accompanying notes in Figure 27. Notice that the student has taken notes selectively, that the notes are generally not taken in complete sentences, and that the student has found it unnecessary to quote any of the material directly and has clearly distinguished his or her own thoughts from those of Darwin.

You probably cannot take notes in your own words if you do not understand what you are reading. Similarly, it is also difficult to be selective in your note-taking until you have achieved a general understanding of the material. I suggest that you first consult at least one general reference text and read the material carefully, as recommended earlier. Once you have located a particularly promising scientific article, read the entire paper through at least once without taking any notes. Resist the (strong) temptation to annotate and take notes during this first reading, even though you may feel that without a pen in your hand you are accomplishing nothing. Put your pencils, pens, and notecards or paper away, and read. Read slowly and with care. Read to understand. Study the illustrations, figure captions, tables, and graphs carefully, and try to develop your own interpretations before reading those of the author(s). Don't be frustrated by not understanding the paper at the first reading; understanding scientific literature takes time and patience . . . and often many rereadings, even for practicing biologists. Concentrate not only on the results reported in the papers but also on the reason the study was undertaken and the way the data were obtained. The results of a study are real; the interpretation of the results is always open to question. And the interpretation is largely influenced by the way the study was conducted. Read with a critical, questioning eye. Many of the interpretations and conclusions in today's journals will be modified in the future. It is difficult, if not impossible, to have the last word in Biology; progress is made by continually building on and modifying the work of others.

By the time you have completed your first reading of the paper, you may find that the article is not really relevant to your topic after all or is of little help in developing your theme. If so, the preliminary read-through will have saved you from wasted note-taking. If you have photocopied the article, all has not been lost; simply turn the pages over and use them for notepaper.

WRITING THE PAPER

Begin by reading all your notes. Again, do this without pen or pencil in hand. Having completed a reading of your notes to get an overview of what you have accomplished, reread them,

Causes of Variability

① WHEN WE COMPARE the individuals of the same variety or sub-variety of our older cultivated plants and animals, one of the first points which strikes us is, that they generally differ more from each other than do the individuals of any one species or variety in a state of nature. And if we reflect on the vast diversity of the plants and animals which have been cultivated, and which have varied during all ages under the most different climates and treatment, we are driven to conclude that this great variability is due to our domestic productions having been raised under conditions of life not so uniform as, and somewhat different from, those to which the parent species had been exposed under nature. There is, also, some probability in the view propounded by Andrew Knight, that this variability may be partly connected with excess of food. It seems clear that organic beings must be exposed during several generations to new conditions to cause any great amount of variation; and that, when the organisation has once begun to vary, it generally continues varying for many generations. No case is on record of a variable organism ceasing to vary under cultivation. ② Our oldest cultivated plants, such as wheat, still yield new varieties: our oldest domesticated animals are still capable of rapid improvement or modification.

Figure 24.   Facsimile of section from Darwin, *The Origin of Species,* from which the notes (Figure 25) on the mechanism of natural selection were taken.

this time with the intention of sorting your ideas into categories. Notes taken on index cards are particularly easy to sort, provided that you have not written many different ideas on a single card; one idea per card is a good rule to follow. To arrange notes written on full-sized sheets of paper, some people suggest annotating the notes with pens of different colors or using a variety of symbols, with each color or symbol representing a particular aspect of the topic. Still other people simply use scissors to snip out sections of the notes, and then group the resulting scraps of paper into piles of related ideas. You should experiment to find a system that works well for you.

   At this point you must eliminate those notes that are irrelevant to the specific topic you have finally decided to write about. No matter how interesting a fact or idea is, it has no place in your paper unless it clearly relates to the rest of the paper and therefore

① Cultivated plants/animals = more variable in appearance than members of any particular species/variety in nature. Greater variation presumed due to greater variation in conditions under which reared.

② All species show variation, even after cultivation by humans for long periods of time; never lose ability to generate new varieties.

③ me: Darwin clearly indicates that individual variation is inherent in living organisms and is central to process of evolution by nat. select., despite no knowledge of genetic mechanism of inheritance.

Figure 25.   Handwritten notes based on a section of Darwin, *The Origin of Species,* concerning natural selection.

helps you develop your argument. Some of the notes you took early on in your exploration of the literature are especially likely to be irrelevant to your essay, since these notes were taken before you had developed a firm focus. Put these irrelevant notes in a safe place for later use; don't let them coax their way into your paper.

You must next decide how best to arrange your categorized notes, so that your essay or term paper progresses toward some conclusion. The direction your paper will take should be clearly and specifically indicated in the opening paragraph, as in the following example written by Student *A*:

Most lamellibranch bivalves are sedentary, living either in soft-substrate burrows (e.g., soft-shell clams, <u>Mya</u> <u>arenaria</u>) or attached to hard substrate

It may be worth while to give another and more complex illustration of the action of natural selection. Certain plants ◄—① excrete sweet juice, apparently for the sake of eliminating something injurious from the sap: this is effected, for instance, by glands at the base of the stipules in some Leguminosæ, and at the backs of the leaves of the common laurel. This juice, though small in quantity, is greedily sought by insects; but their visits do not in any way benefit the plant. Now, let us ◄—② suppose that the juice or nectar was excreted from the inside of the flowers of a certain number of plants of any species. Insects in seeking the nectar would get dusted with pollen, and would often transport it from one flower to another. The flowers of two distinct individuals of the same species would thus get crossed; and the act of crossing, as can be fully proved, gives rise to vigorous seedlings which consequently would have the best chance of flourishing and surviving. The ◄—③ plants which produced flowers with the largest glands or nectaries, excreting most nectar, would oftenest be visited by insects, and would oftenest be crossed; and so in the long-run would gain the upper hand and form a local variety. The flowers, also, which had their stamens and pistils placed, in relation to the size and habits of the particular insects which visited them, so as to favour in any degree the transportal of ④ the pollen, would likewise be favoured. We might have taken the case of insects visiting flowers for the sake of collecting pollen instead of nectar; and as pollen is formed for the sole purpose of fertilisation, its destruction appears to be a simple loss to the plant; yet if a little pollen were carried, at first occasionally and then habitually, by the pollen-devouring insects from flower to flower, and a cross thus effected, although nine-tenths of the pollen were destroyed it might still be a great gain to the plant to be thus robbed; and the individuals which produced more and more pollen, and had larger anthers, would be selected.

Figure 26.   Facsimile of section from Darwin: *The Origin of Species,* from which the notes in Figure 27 were taken.

(e.g., the blue mussel <u>Mytilus</u> <u>edulis</u>)(Barnes, 1980).

However, individuals of a few bivalve species live on the

surface of substrates, unattached, and are capable of

locomoting through the water. One such species is the

scallop <u>Pecten</u> <u>maximus</u>. In this essay, I will explore

the morphological and physiological adaptations that

Plant evolution tied to insect behavior. Flowers now = effective in attracting insects for pollen exchange; how originate?

① Some plant sap apparently toxic (me: no evidence given). Plant nectar makes sap less nasty. Insects attracted to the sweet nectar, even though produced by plant originally to protect plant.

② If plant produces nectar in flower, insects attracted to flower, thus transport pollen, facilitate cross-fert. me: note that selection for nectar prod. in flower can occur only if a few plants accidentally start secreting nectar in flowers. Note that nectar not orig. prod. to attract insects; selected for protect plant, but once being produced, can evolve different function.

③ Those flowers that have greatest success attracting insects will spread the most pollen. me: now know that the genes of these flowers would thus incr. prod. of successful represen- tation in next generation.

④ Nectar prod. not essential to explain evol. of insect-mediated cross pollination. Suppose insect feeds on pollen (as some spp. do). Some pollen would stick to legs and be transferred to another flower. Again, flowers most successful in attracting insects would incr. chance of spreading genes, even though most of the pollen gets eaten.

Figure 27.   Handwritten notes based on another section of *The Origin of Species*.

make swimming possible in P̲. m̲a̲x̲i̲m̲u̲s̲, and will consider
some of the evolutionary pressures that might have selected for these adaptations.

The nature of the problem being addressed is clearly indicated in
this first paragraph, and Student *A* tells us clearly why the problem
is of interest: (1) the typical bivalve doesn't move, and certainly
doesn't swim; (2) a few bivalves can swim; (3) so what is there
about these exceptional species that enables them to do what other
species can't; (4) and why might this swimming ability have evolved?
Note that use of the pronoun *I* is now perfectly acceptable in
scientific writing.

In contrast to the previous example, consider the following
weaker (although not horrible) first paragraph written by Student
B on the same subject:

Most lamellibranches either burrow into, or attach
themselves to, a substrate. In a few species, however,
the individuals lie on the substrate unattached and are
able to swim by expelling water from their mantle cavi-
ties. One such lamellibranch is the scallop P̲e̲c̲t̲e̲n̲ m̲a̲x̲i̲-
m̲u̲s̲. The feature that allows bivalves like P̲. m̲a̲x̲i̲m̲u̲s̲ to
swim is a special formation of the shell valves on their
dorsal sides. This formation and its function will be
described.

In this example, the second sentence weakens the opening paragraph
considerably by prematurely referring to the mechanism of swimming. The main function of the sentence should be to emphasize
that some species are not sedentary; the reader, not yet in a position
to understand the mechanism of swimming, becomes a bit baffled.
The next to last sentence of the paragraph ("The feature that
allows. . .") also hinders the flow of the argument. This sentence
summarizes the essay before it has even been launched, and again,

the reader is not yet in a position to appreciate the information presented; what is this "special formation" and how does it have anything to do with swimming? The first paragraph of a paper should be an introduction, not a summary.

The last sentence of Student *B*'s paragraph does clearly state the objective of the paper, but the reader must ask, "Toward what end?" The author has set the stage for a book report, not an essay. Reread the paragraph written by Student *A* and notice how the same information has been used much more effectively, introducing a thoughtful essay rather than a book report. Student *A*'s first paragraph was written with a clear sense of purpose, with each sentence carrying the reader forward to the final statement of intent. You might guess (correctly, as it turns out) from reading Student *B*'s first paragraph that the rest of the paper was somewhat unfocused and rambling. In contrast, Student *A*'s first paragraph clearly signals that what follows will be well-focused and tightly organized. Get your papers off to an equally strong start.

Another example of a typical, but not especially effective, first paragraph will be helpful:

```
The crustaceans have an exceptional capability for
changing the intensity and pattern of their coloring
(Russell-Hunter, 1979). Many species seem able to change
their color at will. The cells responsible for producing
the characteristic color changes of crustaceans are the
chromatophores. The function of these cells will be dis-
cussed in this essay.
```

What is wrong with this introductory paragraph? The author is certainly off to a strong start with the first sentence. The second sentence, however, begins by repeating information already given in the first sentence (crustaceans can change color), and ends by saying nothing at all (what does "at will" mean for a crustacean?). The last sentence sets up a book report, even though the author calls it an essay. Why will the function of these cells be discussed? More to the point, why should the reader be interested in such a discussion? The reader will be more readily drawn into your intellectual net if you indicate not only where you are heading but also why you are undertaking the journey.

The first paragraph of your paper must state clearly what you are setting out to accomplish and why. Every paragraph that follows the first paragraph should advance your argument clearly and logically towards the stated goal.

State your case, and build it carefully. Use your information and ideas to build an argument, to develop a point, to synthesize. Avoid the tendency to simply summarize papers one by one: They did this, then they did that, and then they suggested the following explanation. Instead, set out to compare, to contrast, to illustrate, to discuss. As in all other scientific writing, always back up your statements with supporting documentation; this documentation may be an example drawn from the literature you have read, or just a reference (author and date of publication) to a paper or group of papers that support your statement, as in the following examples:

```
    There is no evidence for mechanical ventilation in

freshwater pulmonates, so it is presumed that exchange

of gases occurs solely by diffusion (Ghiretti and Ghir-

etti-Magaldi, 1975).


    Schistosomiasis is one of the most serious para-

sitic diseases of mankind, afflicting hundreds of mil-

lions of people and causing hundreds of millions of

dollars in economic losses yearly through livestock in-

festations (Noble and Noble, 1982).


    The ability of an organism to recognize ''self''

from ''non-self'' is found in both vertebrates and in-

vertebrates. Even the most primitive invertebrates show

some form of this immune response. For example, Wilson

(1907) found that disassociated cells from two different
```

species of sponge would regroup according to species;
cells of one species never reaggregated with those of the
second species.

In referring to experiments, don't simply state that a particular
experiment supports some particular hypothesis; describe the relevant
parts of the experiment and explain how the results relate to the
hypothesis under question, as demonstrated in two examples:

> Foreign organisms or particles that are too large to
> be ingested by a single leukocyte are often isolated by
> encapsulation, with the encapsulation response demon-
> strating clear species-specificity. For example, Cheng
> and Galloway (1970) inserted pieces of tissue taken from
> several species into an incision made in the body wall of
> the gastropod <u>Helisoma</u> <u>duryi</u>. Tissue transplanted from
> other species was completely encapsulated within 48
> hours of the transplant. Tissue obtained from individu-
> als of the same species as the host was also encapsu-
> lated, but encapsulation was not completed for at least
> 192 hours.

> Above a certain temperature, further temperature
> increases often have a depressing effect on larval
> growth rates (Kingston, 1974; Leighton, 1974). This
> break point can be very sharply defined. For instance,
> larvae of the bivalve <u>Cardium</u> <u>glaucum</u> were healthy and
> grew rapidly at 31°C, grew abnormally and less rapidly at
> 32–33°C, and grew hardly at all at 34°C (Kingston, 1974).

In all your writing, avoid quotations unless they are absolutely

necessary; use your own words whenever possible. At the end of
your essay, summarize the problem addressed and the major points
you have made, as in the following example:

> In conclusion, the basic molluscan plan for respira-
> tion that had been successfully adapted to terrestrial
> life in one group of gastropods, the terrestrial pulmon-
> ates, has been successfully readapted to life in water by
> the freshwater pulmonates. Having lost the typical mol-
> luscan gills during the evolutionary transition from
> salt water to land, the freshwater pulmonates have
> evolved new respiratory mechanisms involving either the
> storage of an air supply (using the mantle cavity) or a
> means of extracting oxygen while under water, using a gas
> bubble or direct cutaneous respiration. Further studies
> are required for fully understanding the functioning of
> the gas bubble in pulmonate respiration.

Never introduce any new information in your summary paragraph.

CITING SOURCES

Unless you are told otherwise, do not footnote. Instead, cite
references directly in the text by author and date of publication.
For example:

> Survival times of marine organisms exposed to any
> particular pollutant concentration may vary with changes
> in salinity of the test medium, experimental tempera-
> ture, or both (Gray, 1976; Laughlin and Neff, 1979).

Try to make the relevance of the cited reference clear to the reader. For example, rather than writing:

```
Temperature tolerances have been determined for gas-
tropods, bivalves, annelids, and insects (Smith, 1968;
Jones, 1979; Smith and Jones, 1983),
```

it would be clearer to write:

```
Temperature tolerances have been determined for gas-
tropods (Jones, 1979), bivalves and annelids (Smith,
1968), and insects (Smith and Jones, 1983).
```

As always, if you write as though explaining something to yourself, your classmates, or your parents, or if you write so that the paper will be useful to you in the future, you will generally come out ahead.

At the end of your paper, include a section entitled Literature Cited, listing all references you have referred to in your paper. Do not include any references you have not actually read. Each reference listed must give author(s), date of publication, title of article, title of journal, and volume and page numbers. If the reference is a book, the citation must include the publisher, place of publication, and total number of pages in the book, or the page numbers pertinent to the citation. Your instructor may specify a particular format for preparation of this section of your paper. For additional information about preparing the Literature Cited section, see Chapter 2 (pages 66–68).

CREATING A TITLE

By the time you have finished writing, you should be ready to entitle your creation. Give the essay or term paper a title that is appropriate and interesting, one that conveys significant information about the specific topic of your paper (see also p. 64).

No: Plate Tectonics and Mammals
Yes: Evidence for an Influence of Plate Tectonics on the Distribution of Modern Mammals

No: Echinoderm Tube Feet
Yes: The Role of Tube Feet in Echinoderm Locomotion

No: Molluscan Defense Mechanisms
Yes: Behavioral and Chemical Defense Mechanisms of Gastropods and Bivalves

No: Asexual Reproduction: is it the way to go?
Yes: Sexual and Asexual Reproduction in Cnidarians: an evaluation of advantages and disadvantages associated with two different modes of reproduction

REVISING

Once you have a working draft of your paper, you must revise it, clarifying your presentation, expunging ambiguity, eliminating excess words, and improving the logic and flow of ideas. You may also have to edit for grammar and spelling. These topics are considered in detail in Chapter 8.

4

Writing Research Proposals

Research proposals are commonly assigned in advanced Biology courses in place of the more standard "term paper"; the two assignments have much in common, and you should read, or reread, Chapter 3 before proceeding with this chapter. Research proposals, essays, and term papers all involve critical review and synthesis of the primary literature — that is, papers presenting detailed, original results of research rather than articles and books presenting only summaries and interpretations of that research. In addition, however, a research proposal includes a written argument in which you propose to go beyond what you have read; you propose to do a piece of research yourself and seek to convince the reader that what you propose to do should be done and can be done. Research proposals are perhaps the best vehicle for developing your reasoning and writing skills in Biology. This assignment, more than any other, gives you a chance to be creative and to become a genuine participant in the process of biological investigation. Writing a good research proposal is no trivial feat, and the feeling of accomplishment that descends upon you once you are finished is indescribably nice.

Research proposals have two major parts: a review of the relevant scientific literature and a description of the proposed research. In the first part, you review the primary literature on a particular topic, but you do so with a particular goal in mind: You wish to lead your reader to the inescapable conclusion that the question you propose to address follows logically from the research that has gone before. Writing a research proposal rather than a term paper thus helps you avoid falling into the "book report" trap; once you develop a research question to ask, you should have an easier time focusing your literature review on the development of a single, clearly articulated theme.

In addition to providing you with a convenient vehicle for exploring and digesting the primary scientific literature and for focusing your discussion of that literature, you may find that the research you propose to do is actually do-able — and do-able by you. Your proposal could turn out to be the basis for your own summer research project, senior thesis, or even master's or Ph.D. thesis.

RESEARCHING YOUR TOPIC

Proceed as you would for researching a term paper or essay (see Chapter 3). For this assignment especially, it will be important to have a firm grasp of your subject before you plunge into the original, primary scientific literature, so read the appropriate sections of several recent general textbooks before you look elsewhere. The next step should be to browse through recent issues of appropriate scientific journals; your instructor can suggest several that are particularly appropriate to your topic of interest.

Before you roll up your sleeves and prepare to wrestle in earnest with a published scientific paper, read it through once for general orientation. Once you begin your second reading of the paper, don't allow yourself to skip over any sentences or paragraphs you don't understand. Keep a relevant textbook by your side as you read the primary literature so that you can look up unfamiliar facts and terminology.

I mentioned previously that the results of a study depend largely on the way the study was conducted. We have also seen that although the results of a study are real, the interpretation of those results is always subject to change. The Materials and Methods section and the Results section of research papers must therefore be read with particular care and attention. Scrutinize every table and graph until you can reach your own tentative conclusions about the results of the study before allowing yourself to be swayed by the author's interpretations. Read with a questioning, critical eye.

The primary scientific literature must be read slowly, thoughtfully, and patiently, and a single paper must usually be reread several times before it can be thoroughly understood; don't become discouraged after only one or two readings. Reading the scientific literature is slow going, but, like playing tennis or sight-reading music, it gets easier with practice. If, after several rereadings of the paper you are hovering over, and if, after consulting several textbooks, you are still baffled by something in the paper you are reading, ask your instructor for help.

As you carefully read each paper, pay special attention to the following:

1. What specific question is being asked?
2. How does the design of the study address the question posed?
3. What are the controls for each experiment?
4. How convincing are the results? Are any of the results surprising?
5. What contribution does this study make toward answering the original question?
6. What aspects of the original question remain unanswered?

Reread the paper until you can answer each of these questions. Then ask yourself an additional question:

7. What might be a next logical question to ask, and how might this question be addressed?

Continue your library research using the references listed at the end of the recent papers you are reading, and perhaps by consulting *Biological Abstracts* or one of the other indexing services discussed in Chapter 3. One particularly convenient thing about preparing a research proposal is that it's relatively easy to tell when your library work is finished; it's finished when you know what your proposed research question will be, and when you know exactly why you are asking that question.

WRITING THE PROPOSAL

Divide your paper into three main portions: Introduction, Background, and Proposed Research.

Introduction

Give a brief overview of the research being considered and an indication of the nature of the question you will pursue, as in the following example:

> Studies have shown that such endurance exercises as running and swimming can affect the reproductive physiology of women athletes. Female runners (Dale et al., 1979; Wakal et al., 1982), swimmers (Frisch et al., 1981), and ballet dancers (Warren, 1980) menstruate infrequently (oligomenorrhea) in comparison with nonathletic women of comparable age, or not at all (amenorrhea). The degree of menstrual abnormality varies directly with the intensity of the exercise. For example, Malina et al. (1978) have shown that menstrual irregularity is more common, and more severe, among tennis players than among golfers.
>
> The physiological mechanism through which strenuous activity disrupts the normal menstrual cycle is not yet clear; inadequate fat levels (Frisch et al., 1981), altered hormonal balance (Sutton et al., 1973), and physiological predisposition (Wakat et al., 1982) have each been implicated.
>
> In the proposed research, I will study 175 women weight lifters in an attempt to determine the relative

```
importance of fat levels, hormone levels, and physiolog-
ical predisposition in promoting oligomenorrhea and
amenorrhea.
```

Notice that the author of this proposal has not used the Introduction to discuss the question being addressed, or to describe how the study will be done. The Introduction provides only (1) general background to help the reader understand why the topic is of interest, and (2) a brief but clear statement of the specific research topic that will be addressed. A detailed discussion of prior research belongs in the Background section of the proposal, and a detailed description of the proposed study belongs in the Proposed Research section of the proposal.

Notice also that every factual statement (for example, "Female runners . . . menstruate infrequently") is supported by a reference to one or more papers from the primary literature. These references enable the reader to obtain, painlessly, additional information on particular aspects of the subject and to verify the accuracy of statements made in the proposal. Backing up statements with references also protects the author of the proposal by documenting the source of information; if the author of your source is mistaken, why should you take the blame?

Background

Discuss the relevant literature, leading up to the specific objective of your proposed research. This section of your proposal follows the format of a good term paper or essay, as already described in Chapter 3 (pp. 99–108). In a proposal, however, the Background section will end with a brief statement of what is now known and what is not yet known about the research topic under consideration, and a clear, specific description of the research question(s) you propose to investigate. Here are two examples. The author of the first example has already spent two and a half pages of the Background section describing documented effects of organic pollutants on adults and developmental stages of various marine vertebrates and invertebrates.

Thus many fish, echinoderm, polychaete, mollusc, and crustacean species are highly sensitive to a variety of fuel oil hydrocarbon pollutants, and the early stages of development are especially susceptible. However, many of these species begin their lives within potentially protective extra-embryonic egg membranes, jelly masses, or egg capsules (Anderson et al., 1977; Eldridge et al., 1977; Kînéhcep, 1979). The ability of these structures to protect developing embryos against water-soluble toxic hydrocarbons has apparently never been assessed.

The egg capsules of marine snails are particularly complex, both structurally and chemically (Fretter, 1941; Bayne, 1968; Hunt, 1971). Such capsules are typically several mm to several cm in height, and the capsule walls are commonly 50-100 μm[1] thick (Hancock, 1956; Tamarin and Carriker, 1968). Depending on the species, embryos may spend from several days to many weeks developing within these egg capsules prior to emergence as free-swimming larvae or crawling juveniles (Thorson, 1946).

Little is known about the tolerance of encapsulated embryos to environmental stress, or about the permeability of the capsule walls to water and solute. Kînéhcep (1982, 1983) has found that the egg capsules of several shallow-water marine snails (Ilyanassa obsoleta, Nucella lamellosa, and N. lapillus) are permeable to both salts and water, but are far less permeable to the small organic molecule glucose. Capsules of at least these

[1] μm = microns (10^{-6} meters).

species are thus likely to protect embryos from exposure
to many fuel oil components.

In the proposed study, I will (1) document the tolerance of early embryos of N. lamellosa and N. lapillus,
both within capsules and removed from capsules, to the
water soluble fraction of Number 2 fuel oil; (2) determine the general permeability characteristics of the
capsules of these two gastropod species to see which
classes of toxic substances might be unable to penetrate
the capsule wall; and (3) use radioisotopes to directly
measure the permeability of the capsules to several major components of fuel oil.

The second, shorter example concerns the hormonal control
of reproductive activity in sea stars. The author of this proposal
has already spent three pages of the Background section discussing
experiments demonstrating that (1) gamete release (spawning) is
under hormonal control; (2) the response to the hormone varies
seasonally; and (3) the variation in response seems to reflect changing
concentrations of an inhibitory hormone called shedhibin.

. . . The present evidence suggests, therefore, that
the influence of the excitatory hormone is regulated by
seasonal fluctuations in the secretion of shedhibin, although seasonal changes in the concentrations of this
inhibitory hormone have not yet been documented.

In the proposed research, I will:

(1) identify the time of year during which gamete release is inhibited in mature sea stars (Asterias
forbesi);

(2) develop a monoclonal antibody to the inhibitory
substance shedhibin;

(3) and use immunofluorescent techniques to quantify
the amount of shedhibin produced and secreted at different stages of the reproductive cycle of <u>A</u>. <u>forbesi</u>.

Proposed Research

This portion of your proposal has two interrelated components:

1. What specific question(s) will you ask, and
2. How will you address each of these questions?

Different instructors will put different amounts of stress on Parts
1 and 2. For some of us, the formulation of a valid and logically
developed question is the major purpose of the assignment, and
a highly detailed description of the methodology will not be required.
For such an instructor, you may, for example, propose to extract
and separate proteins without actually having to know in detail
how this is accomplished. But other instructors may feel that your
mastery/knowledge of methodological detail is as important as
the validity of the questions posed. Both approaches are defensible,
depending in large part on the nature of the field of inquiry, on
the level of the course being taken, and on the amount of laboratory
experience you have had. Be sure you understand what your instructor expects of you before preparing this section of your paper.

As you describe each component of your proposed research,
indicate clearly what specific question each experiment is designed
to address, as in the following examples.

To see if there is a seasonal difference in the
amount of hormone present in the bag cells that induce
egg-laying in <u>Aplysia</u> <u>californica</u>, bag cells will be
dissected out of mature individuals each month and . . .

Before the influence of light intensity on the rate
of photosynthesis can be documented, populations of the

test species (wild columbine, <u>Aquilegia</u> <u>canadensis</u>) must
be established in the laboratory. This will be done
by . . .

To monitor seasonal changes in the relative abundance of macroalgae at different levels in the rocky intertidal zone at Nahant, Mass., I will inspect each of
the ten boulders at monthly intervals for a twelve-month
period. At each inspection, I will . . .

CITING REFERENCES AND PREPARING THE LITERATURE CITED SECTION

References are cited directly in the text by author and year,
as in the examples given above (see also pp. 5, 106, 108–109).
The Literature Cited section of your proposal is prepared as already
described for term papers and essays (see Chapter 2, pp. 66–68).

5

Writing Summaries and Critiques

For assignments in writing summaries and critiques, you are asked to read a paper from the original scientific literature and summarize or assess that paper, usually in fewer than two double-spaced, typewritten pages. *Brief* does not, in this case, mean *easy*. In fact, producing that one- or two-page summary or critique will probably require as much mental effort as that involved in preparing an essay or term paper of from five to ten pages in length. To do well in these short assignments you must fully understand what you have read, which usually means that you must read the paper many times, slowly and thoughtfully.

Follow the same procedures whether you are asked to write a summary or a critique; indeed, a critique begins as a summary, to which you then add your own evaluation of the paper.

To begin, read the paper once or twice without taking notes. Fight the temptation to underline, highlight, or otherwise create the illusion that you are accomplishing something. It is often difficult to distinguish the significant from the not-so-significant points during the first reading of a scientific paper; skim the paper once for general orientation and overview. Don't try for detailed understanding in the first reading, but do jot down any unfamiliar terms or the names of unfamiliar techniques so that you can look these up in a textbook before you reread the paper. It often helps to consult a textbook about the general biology of the organisms being studied before returning to the paper.

During the next, more careful reading of the paper, pay special attention to the Materials and Methods and the Results sections; the essence of any scientific paper is contained here. The results obtained in a study depend on the way the study was conducted. Were samples taken only at one particular time of year? Was the study replicated? How many individuals were examined? What techniques were used? In an experiment, what variables (for example, photoperiod, temperature, salinity, or food supply) were held constant? Were proper controls provided for each experiment? Which factors might affect the outcome of the study?

As you begin to study the Results section, scrutinize every graph, table, and illustration, developing your own interpretations of the data before rereading the author's verbal presentation. We are readily influenced by the opinions of others, especially when those opinions are well-written. Keep an open mind when reading the author's words, but try to form your own opinions about the data first; you may see something that the author did not.

WRITING THE FIRST DRAFT

You will know that you are ready to write your first draft of the assignment when you can distill the essence of the paper into a single, intoxicating summary sentence, or, at most, two summary sentences. As a general rule, do not begin to write your review until you can write such an abbreviated summary; this exercise will help you discriminate between the essential points of the paper and the extra, complementary details. Several examples of good summary sentences are given below.

If you cannot write a satisfactory one- or two-sentence summary, reread the article; you'll get it eventually. Once your summary sentence is committed to paper, ask yourself these questions:

1. Why was the study undertaken? To answer this, draw especially from information given in the Introduction and Discussion sections of the paper you have read.
2. What specific questions were addressed? Summarize each question in a single sentence.
3. How were these questions addressed? What specific approaches were taken to address each question on your list?
4. What were the major findings of the study?
5. What questions remain unanswered by the study? These may

be questions addressed by the study but not answered conclusively, or they may be new questions arising from the
findings of the study under consideration.

WRITING THE FINAL DRAFT

When you can answer these questions without referring to
the paper you have read, you can begin to write.

At the top of the page, beginning at the left-hand margin,
give the complete citation for the paper being discussed: names
of all authors, year of publication, title of the paper, title of the
journal in which the paper was published, and volume and page
numbers of the article. On a new line, indent five spaces and begin
your summary or critique with a few sentences of background
information. Your introductory sentences must lead up to a statement
of the specific questions the researchers set out to address. Next,
tell (1) what approaches were used to investigate each question
and (2) what major results were obtained. Be sure to state, as
succinctly as possible, exactly what was learned from the study.

When writing a critique, you get to insert your own ideas
about the paper you have read. This does not mean you should
set out to tear the paper to shreds; a critical review is a thoughtful
summary and analysis, not an exercise in character assassination.
Almost every piece of biological research has shortcomings, most
of which become obvious only in hindsight. Yet every piece of
research contributes some information, even when the original
goals of the study are not attained. Emphasize the positive — focus
on what *was* learned from the study. Although you should not
dwell on the limitations of the study, you should point out these
limitations toward the end of your critique. Were the conclusions
reached by the authors out of line with the data presented? Do
the authors generalize far beyond the populations or species studied?
Which questions remain unanswered? How might these questions
be addressed? How might the study be improved or expanded in
the future? Keep this in mind as you write: You wish to demonstrate
to your instructor (and to yourself) that you understand what you
have read. Do not comment on whether or not you enjoyed the
paper, or found it to be well-written; stick to the science.

To cover so much ground within the limits of one typewritten page is no small feat, but it can be done if you first make certain that you fully understand what you have read. Consider the following example of a brief, successful summary. Before writing the summary the student condensed the paper into these two sentences:

The tolerance of a Norwegian beetle (<u>Phyllodecta</u> <u>la-</u><u>ticollis</u>) to freezing temperatures varies seasonally, in association with changes in the blood concentration of glycerol, amino acids, and total dissolved solute. However, the concentration of nucleating agents in the blood does not vary seasonally.

THE SUMMARY

Minnie Leggs
Bio 101
Fall 1984

Van der Laak, S. 1982. Physiological adaptations to low temperature in freezing-tolerant <u>Phyllodecta</u> <u>laticollis</u> beetles. <u>Comp</u>. <u>Biochem</u>. <u>Physiol</u>. 73A: 613—620.

Adult beetles (<u>Phyllodecta</u> <u>laticollis</u>), found in Norway, are exposed to sub-zero (°C) temperatures in the field throughout the year. In general, organisms that tolerate freezing conditions either produce nucleating agents that trigger ice formation outside the cells rather than within them or they produce biological anti-freezes, such as glycerol, that lower the freezing point of the blood and tissues to below that of the environment, thereby preventing ice formation. This study was undertaken to document the tolerance of <u>P</u>. <u>laticollis</u> to below-freezing temperatures and to account for seasonal shifts in the temperature tolerance of these beetles.

Beetles were collected throughout the year and frozen to temperatures as low as −50°C; post-thaw survivorship was then determined. Determinations were also made of the concentrations of solutes in the blood (that is, blood osmotic concentration), total water content, amino acid and glycerol concentrations in the blood, presence of nucleating agents in the blood, and the temperature to which blood could be super-cooled before freezing would occur.

The temperature tolerance of P. laticollis varied from about −9°C in summer to about −42°C in winter; this shift in freezing tolerance was paralleled by a dramatic winter increase in glycerol concentration and in total blood osmotic concentration. Amino acid concentration also increased in winter, but the contribution to blood osmolarity was small compared to that of glycerol. Nucleating agents were present in the blood year-round, ensuring that ice formation will occur extracellularly rather than intracellularly, even in summer.

For beetles collected in mid-winter and early spring, blood glycerol concentrations could be artificially reduced by warming beetles to 23°C (room temperature) for about 24–150 h. When glycerol concentrations of spring and winter beetles were reduced to identical levels by warming, the spring beetles tolerated freezing better than the winter beetles; these differences in tolerance could not be explained by differences in amino acid concentrations. This result indicates that some other factors, as yet unknown, are also involved in determining the freezing tolerance of these beetles.

Analysis of Student Summary

The student has within one typed page successfully distilled a seven-page technical report to its scientific essence. Note that the student used the first three sentences to introduce the topic and then summarized the purpose of the research in one sentence. The next short paragraph summarizes the experimental approach taken, and the main findings of the study are then stated. No superfluous information is given; the author of this assignment provided only enough detail to make the summary comprehensible. The product glistens with understanding. Rereading the student's two-sentence encapsulation of the paper (p. 123), you can see that the student was indeed ready to write the report.

As the following example shows, a critique is very much like a summary, except that the student adds his or her own assessment at the end. Before writing the critique, the student produced this one-sentence summary of the paper:

> The egg capsules of the marine snails <u>Nucella</u> <u>lamellosa</u> and <u>N</u>. <u>lima</u> protect developing embryos against low-salinity stress, even though the solute concentration within the capsules falls to near that of the surrounding water within about one h.

THE CRITIQUE

Saul Tee
Bio 101
Fall 1985

Kînehcép, N. A. 1982. Ability of some gastropod egg capsules to protect against low-salinity stress. <u>J</u>. <u>Exp</u>. <u>Marine</u> <u>Biol</u>. <u>Ecol</u>. 63: 195–208.

The fertilized eggs of marine snails are often enclosed in complex, leathery egg capsules with 30 or more embryos being confined within each capsule. The embryos develop for one or more weeks before leaving the cap-

sules. The egg capsules of intertidal species potentially expose the developing embryos to thermal stress,
osmotic stress, and desiccation stress. This paper describes the ability of such egg capsules to protect developing embryos from low-salinity stress, such as might
be experienced at low tide during a rainstorm.

Two snail species were studied: <u>Nucella</u> <u>lamellosa</u>
and <u>N</u>. <u>lima</u>. Embryos were exposed, at 10-12°C, either to
full-strength seawater (control conditions) or to 10-12%
seawater solutions (seawater diluted with distilled
water). The ability of egg capsules to protect the enclosed embryos from low salinity stress was assessed by
placing intact egg capsules into the test solutions for
up to 9 h, returning the capsules to full-strength seawater, and comparing subsequent embryonic mortality with
that shown by embryos removed from capsules and exposed
to the low-salinity stress directly.

Encapsulated embryos exposed to the low salinities
suffered less than 2% mortality, even after low-salinity
exposures of 9 h duration. In contrast, embryos exposed
directly to the same test conditions for as little as 5 h
suffered 100% mortality. All embryos survived exposure
to control conditions for the full 9 h, showing that removal from the capsules was not the stress killing the
embryos in the other treatments. Sampling capsular fluid
at various times after capsules were transferred to the
diluted seawater, Kînehcép found that the concentration
of solutes within capsules fell to near that of the surrounding water within about one h after transfer.

This study clearly demonstrates the protective value of the egg capsules of two snail species faced with low-salinity stress. However, Kînehcep was unable to explain how egg capsules of these two species protect the enclosed embryos, since the capsules do not prevent decreases in the solute concentration of the capsular fluid. Although Kînehcép plots the rate at which the solute concentration falls within the capsules (his Fig. 1), he sampled only at 0, 60, and 90 minutes after the capsules were transferred to water of reduced salinity. I think he should have sampled at frequent intervals during the first 60 min to discover how rapidly the solute concentration of the capsule fluid falls. As Kînehcép himself suggests, perhaps the embryos are less stressed if the concentration inside the capsule falls slowly.

These experiments were all performed at a single temperature even though encapsulated embryos are likely to experience fluctuation in both temperature and salinity as the tide rises and falls during the day; the study should be repeated using a range of temperatures likely to be experienced in the field. In addition, I suggest repeating these experiments using deep-water species whose egg capsules are never exposed to salinity fluctuations of the magnitude used in this study.

Analysis of Student Critique

Again, this student begins with just enough introductory information to make the point of the study clear and ends the first paragraph with a succinct statement of the researcher's goal. The methods and results of the study are then briefly reviewed, as in a summary. Whereas a summary would probably end at this point, the critique continues with thought-provoking assessments by the student. Note that the student was careful to distinguish his thoughts from those of the paper's author (see pp. 94–95, on plagiarism).

Clearly, successful completion of either type of assignment is no trivial matter. But preparing good summaries and critiques is an excellent way to push yourself toward true understanding of what you read, and of the nature of scientific inquiry.

6

Preparing a Paper Presentation

Oral in-class presentations of published research papers are often assigned in conjunction with or in place of written summaries or critiques. I include this short chapter in a book about writing because oral presentations are developed in much the same way as written summaries or critiques. The goals are virtually identical in the two cases: In both oral and written assignments, you seek to capture the essence of the research project — why it was undertaken, how it was undertaken, and what was learned. Effective speakers often prepare detailed notes for their talks, even though they do not necessarily refer extensively to these notes during the presentation. Indeed, the notes for a good oral presentation can provide an ideal framework for later expansion into a written paper of any size. In writing a paper, it is often helpful to think first in terms of giving an oral presentation.

Despite some major similarities, an oral presentation must differ from a written presentation in one important respect: A typewritten page can be read slowly and pondered, and can be reread as often as necessary, until all points are understood; an oral report, however, gives the listener only one chance to grasp the material. An analogy can be made with music. Before about 1910, music, to be successful, had to be liked at the first hearing; appreciation of subtleties might grow with repeated listenings, but composers knew that if their audience was not captivated by the first performance, that performance might well be the last. It was only with the invention of the phonograph that composers could survive while delivering music intended to grow on its audience. An oral presentation goes past the listener only once: For maximum impact, it must be very well organized, developed logically, and stripped of details that divert the listener's attention from the essential points of the presentation.

Like a good summary or critique, your talk can be effective only if you fully understand your topic. As suggested in the preceding chapter, it is wise to skim the paper once or twice for general orientation, consult appropriate textbooks for background information as necessary, and pay particular attention to the Materials and Methods section and to the tables, graphs, and photographs included in the Results section of the paper. When you can summarize the essence of the paper in one or two sentences, you are ready to prepare your talk.

PREPARING THE TALK

1. If you wish to keep your audience awake, do not simply paraphrase the Introduction, Materials and Methods, Results, and Discussion sections of the paper. If you intend to make an effective presentation, you must reorganize the information in the assigned paper. Begin your talk by providing background information, drawing from the Introduction and Discussion sections of the paper and from outside sources if necessary, so that the listener can appreciate why the study was undertaken. End your introduction with a concise statement of the specific question or questions addressed in the paper.

2. Be selective; delete extraneous details. Much of what is appropriate in a research paper is not necessarily appropriate for a talk about that paper. Since the listener has only one chance to get the point, some of the details in the paper must be pruned out in preparing the oral presentation. Streamline: Include only those details needed to understand what comes later. If, for example, you will not discuss the influence of sample size on the results obtained, do not burden the listener with such details in your talk.

Similarly, is it important that the samples were mixed on a shaker table? If you never discuss this detail later in your talk, omit it at the outset. Tidbits such as these sometimes come out in the question period following your presentation but do not belong in the presentation itself.

3. Focus your talk on the results.

4. Draw conclusions as you present each component of the study, so that you lead in logical fashion to the next question addressed. For example: "The oyster larvae grew 20 μm/day when fed diet *A*, 25 μm/day when fed diet *B*, and 65 μm/day when fed a combination of diets *A* and *B*. This suggests that important nutrients missing in each individual diet were provided when the diets were used in combination. To determine what these missing nutrients might be, ..." Lead your audience by the nose from point to point.

5. Plan to use the blackboard. A simple summary table or two is helpful when numbers are being discussed; numbers floating around in the air are difficult for listeners to keep track of. A diagram of experimental protocol can help the listener follow the plan of a study. Data can often be effectively summarized in a few graphs, even when those data were presented in the original paper as complicated tables. Keep the graphs simple, and be sure to label both axes. You need not reproduce graphs exactly as given in the paper, and you need not display every entry from a particular table. Focus on showing the trends in the data, and omit anything that fails to help you make your point clearly.

6. Summarize the major findings of the research at the end of your talk, driving the points home one by one. You may wish to end your talk with a brief discussion of the way the study could be improved or expanded in the future, but don't set out to discredit the authors. End on a positive note, reinforcing what you want your audience to remember.

7. Be prepared for questions about methodology. Listeners often ask about interpretations of the data; to answer these questions, you must be thoroughly familiar with the way the study was conducted.

GIVING THE TALK

1. Know what you're going to say and how you're going to say it. Hesitation, vagueness, and searching for words will all suggest a lack of understanding and, in addition, will lose the attention of your audience. Write out your talk and practice it until you can produce a smooth delivery.

2. Write on the blackboard during your presentation, as, for example, when labelling the axes of graphs. This helps punctuate your statements and also gives the listener time to digest what you are showing as well as time to take notes. For the same reason, you should label curves as you draw and talk about them. It is often a mistake to put completed illustrations on the blackboard ahead of time; the listener generally gets deprived of the opportunity to absorb what is being presented.

3. Make the blackboard work for you by drawing the listeners' attention to specific aspects of the graphs and tables that represent the point you wish to make. Don't simply say, "This is shown in the graph on the board." Rather, say "For example, all the animals fed on diets A and B grew at comparable rates, but . . . ," and be sure to point to the data as you speak.

4. Write unfamiliar terms on the blackboard.

5. Don't mumble. Don't read your talk. Don't talk to the blackboard. Make eye contact with your listeners.

6. Don't automatically refer to the author of a paper as *he*. Many papers are written by women, and many are written by two or more researchers.

7. Don't end abruptly. Warn your audience when you are nearing the end of your talk by saying something like, "I would like to make one final point," or "Before I end, I wish to emphasize that . . ." Such phrasings will prepare the listeners to receive your summary statements.

8. End your talk gracefully. A self-conscious giggle or a "Well, I guess that's it" isn't the best way to close an otherwise captivating presentation. I suggest something like, "Are there any questions?"

9. Do not allow your presentation to exceed the time allotted. You will lose considerable good will by rambling on beyond your time limit. Here again, a few practice sessions come in handy.

10. During the question period after your presentation has been given, do not feel compelled to answer questions that you don't understand; politely ask for clarification until you figure out what is being asked.

11. Do not be afraid to admit that you don't know the answer to a question. You can easily work your neck into a noose by pretending you know more than you really do; nobody expects you to be the world's authority on the topic you are presenting. Simply saying, "I don't know" is the safest way to go.

7
Writing Letters of Application

An application for a job, or for admission to a graduate or professional program, will generally include a résumé and accompanying cover letter (both of which you write), and several letters of recommendation (which you generally never get to see). When applying to graduate or professional schools, and often when applying for jobs, you will also include a transcript of your college coursework and any special examination scores — for example, Graduate Record Examination (GRE) scores. You have no control over what your transcript and GRE scores now say about you; what is done is done. But you can still influence the message transmitted through your résumé and supporting letters.

Your résumé summarizes your educational background, relevant work experience, relevant research experience, goals, and general interests. The accompanying cover letter identifies the position for which you are applying and draws the reader's attention to those aspects of your résumé that make you a particularly worthy candidate. The recommendations will give an honest assessment of your strengths and weaknesses (we all have some of each) and offer the reader an image of you as a person and as a potential employee or participant in a professional program. In this chapter, I will consider the art of preparing effective résumés and cover letters, and of increasing the odds of ending up with effective letters of recommendation.

BEFORE YOU START

Always try to put yourself in the position of the people who will be reading your application. What will they be looking for? They will probably be considering your application with three main questions in mind:

1. Is the applicant qualified for this particular position?
2. Is the applicant really interested in our program or company?
3. Will the applicant fit in here?

Your application must address all three issues.

When you prepare your application, you should also consider that the number of applications received by a potential employer or professional school usually exceeds the number of positions available, and often by a considerable margin. Many applicants will be qualified for the position, yet not every applicant can be interviewed or offered admission. Whoever begins reading your application will necessarily be looking for *any* excuse to disqualify you from the competition; your goal, then, must be to hold the reader's interest to the end.

PREPARING THE RÉSUMÉ

The people who read your application will not spend hours scrutinizing your résumé; probably they will examine it for only one or two minutes at most. An effective résumé is therefore well-organized, neat, and as brief as possible. Your résumé should not exceed two pages in length.

There is no standard format for a résumé; the model given in Figure 28 should be modified in any way that emphasizes your particular strengths and satisfies your own esthetic sense. The résumé is, however, no place to be artsy or cute; don't do anything that might suggest that you are not taking the application process seriously.

Steven W. Ross

Address and phone number
Until June 1, 1986:
P.O. Box 272
Cambridge, MA 02138
(617) 201-1717

After June 1:
3730 Vernon Place
Raleigh, NC 27607
(919) 788-0153

Date of Birth: May 5, 1965

Goals: To earn a Ph.D. in Comparative Physiology and pursue a career in teaching and research.

Education

Tufts University, Fall 1982–Spring 1986
Major: Biology (B.S., cum laude)

Research Experience

Conducted a one-semester research project on the structure and function of guard cells in lyre-leaved sage, <u>Salvia</u> <u>lyrata</u> L., using transmission electron microscopy. Presented the results of this research at the 29th New England Undergraduate Research Conference, Framingham, MA, May 1986.

Teaching Experience

Undergraduate teaching assistant for introductory Biology laboratory, Fall semesters 1984 and 1985

Honors

Elected to Phi Beta Kappa honor society, Spring 1986
Received Churchill Prize in Biology (for performance
in introductory Biology course), Spring 1983
Dean's List seven out of eight semesters
Selected for teaching assistant position noted above

Work Experience

Summer 1981, 1982. Counselor, Lake Baker Summer
Camp, Alaska

Figure 28.   Sample Résumé

    Summer 1983, 1984. Worked for U.S. Post Office, sort-
      ing and delivering mail
    Summer 1985. Construction work, Ahab's Boat Yard,
      Nantucket, MA

Special Skills

    Tissue preparation (fixation, embedding, section-
      ing) for transmission electron microscopy
    Operation of RCA—EMU3 transmission electron
      microscope
    Developing 35mm black and white film and TEM nega-
      tives (glass plates)
    Printing black and white photographs

Outside Activities

    Saxophonist, Tufts Jazz Band (1983—1986)
    Swim team (1982—1985, captain 1985)
    Campus tour guide (1985—1986)

Figure 28.   Sample Résumé *(continued)*

All résumés must contain the following three components:

1. full name, address, telephone number;
2. educational history;
3. relevant work, teaching, and research experience, if any.

In addition, you will want to add any other information that makes you look talented or well-rounded or both:

4. honors received;
5. papers published;
6. special skills;
7. outside activities, sports, hobbies.

Avoid drawing attention here to any potential weaknesses; if, for example, you lack teaching experience, do not write "Teaching experience: none." Use the résumé exclusively to play up your strengths.

You might also add a one- or two-sentence statement of your immediate and long-range goals, if known, and the names of people who have agreed to write references on your behalf; this material is often incorporated into the cover letter instead.

You are not required to list age, race, marital status, height, weight, sex, or any other personal characteristic. Be self-serving in deciding what to include. If you think your youth might put you at a disadvantage, omit this information. If you think your age, race, or sex might give you a slight competitive edge, by all means include it.

Do not be concerned if your first résumé looks skimpy; it will fill out as the years go by. It is better to present a short, concise résumé than an obviously padded one.

You should alter your résumé for each application completed, to focus on the different strengths required by different jobs or programs. If, for example, Steven W. Ross, whose résumé appears in Figure 28, were applying to a marine underwater research program, he might add, under Special Skills, that he is a certified SCUBA diver. If he were applying for a laboratory job in a hospital, he would probably omit the information about diving certification.

PREPARING THE COVER LETTER

The cover letter plays a large role in the process of application and is usually the first part of your application read by an admissions committee or prospective employer. A well-crafted letter of application can do much to counteract a mediocre academic record. A poorly crafted letter, on the other hand, can do much to annihilate the good impression made by a strong academic performance. Keep revising this letter until you know it works well on your behalf. Have some friends, or perhaps an instructor, read and comment on your letter; then revise it again. Be sure to type the final copy; neatness counts, and typing also conveys seriousness of purpose. The time you put into polishing your cover letter is time well spent. The cover letter should be about one or, at most, two typed pages.

Do not simply write,

Dear M. Pasteur:

I am applying for the position advertised in the <u>Boston Globe</u>. My résumé is enclosed. Thank you for your consideration.

                              Sincerely,

                              Earl N. Meyer

                              Earl N. Meyer

Although the letter ends well, its beginning is vague and its midsection does little to further the applicant's cause. Use the cover letter to:

1. Identify the specific position for which you are applying (Monsieur Pasteur may have several positions open. Earl N. Meyer is applying for the position of research assistant, but how is M. Pasteur to know?);
2. draw the reader's attention to those elements of your résumé that you feel make you a particularly qualified candidate;
3. indicate that you understand what the position entails, and that you have the skills necessary to do a good job;
4. convince the reader you are a mature, responsible person;
5. convey a genuine sense of enthusiasm and motivation.

Before you begin to write the letter, ask yourself some difficult questions and jot down some carefully considered answers:

Why do I want this particular job or to enter this particular graduate program?

What skills would be most useful in such a job or program?

Which of these skills do I have?

What evidence of these skills can I present?

Your answers to these questions will provide the pattern and the yarn from which you will weave your cover letter.

Not everyone will have a résumé that looks like Steven Ross's (Figure 28). But you need not have made the Dean's List every semester or have had formal teaching or research experience in order to impress someone with your application. In your cover letter, focus on the experiences that you *have* had. In lieu of teaching, perhaps you have done formal or informal tutoring. In lieu of having had formal research experience, perhaps you have taken numerous laboratory courses. Perhaps some experience you had in one or more of these laboratory courses influenced your decision to apply to a particular job or program. Perhaps acquiring certain skills in one or more of these laboratory courses has prepared you for the program or position to which you are applying. Or perhaps you can draw from experience outside Biology to document reliability, desire and willingness to learn new things, or ability to learn new techniques quickly. We all have strengths; decide what yours are and which ones are appropriate for inclusion in your application.

Tailor each letter to the particular position or program for which it is being prepared. Try to find some special reason for applying to each program; if possible, your application should reflect deliberate choice and a clear sense of purpose. If, for example, you have read papers written by a faculty member at the institution to which you are applying and have become interested in that person's research, weave this information into your cover letter. Should you take this approach, you must say enough about that person's research or research area to make clear that you understand what you are writing about. On the other hand, if your major reason for wanting a particular job or wanting to attend a particular graduate program is the geographic location of the company or school, be careful not to state this as your sole reason for applying; as you write, and as you reread what you have written, try to put yourself on the receiving end of the cover letter and consider how your statements might be interpreted.

Back up all statements with supporting details. Avoid simply saying that you have considerable research experience. Instead, briefly explain what your research experience has been. Do not state that you are a gifted teacher; describe your teaching experience. State the facts and let the reader draw the proper inferences.

Sign the letter with your given name, not a nickname; again, don't run the risk of not being taken seriously.

Here is an example of a weak cover letter. Similar letters

have, unfortunately, been submitted by people with very good grades, test scores, and letters of recommendation. The author is applying for admission to a Ph.D. program in Biology.

To the admissions committee:

I have always been fascinated by the living world around me. I marvel at the details of the way Biology works, and I would now like to fulfill my curiosity and passion for Biology in pursuing a Ph.D. in your program.

As you can see from my transcript, I have taken twelve courses in Biology (two more than the number needed for graduation) and have done well in most of them. I am especially interested in plant physiology and did a one-semester research project on this subject during my senior year.

I would also like to apply for the teaching assistantship award. I have always liked helping people learn about science, and I am eager to communicate my enthusiasm for Biology to others. I have requested that my GRE scores be sent directly to you.

I look forward to your reply.

Sincerely,

Dineen Arms

Dineen Arms

Remember, the admissions committee is looking for any excuse to disqualify an applicant; this letter may give the committee just that excuse, regardless of what the rest of the application looks like. The letter does convey enthusiasm, but it is a very naive enthusiasm. What does the student find interesting about the physiology of plants? What was the research project? What question was asked? How was the question addressed? What results were obtained? Did the student learn anything from the experience? Does the student really know anything about plant physiology? What makes the student think that she would be an effective teacher? Has she had any teaching experience? Does she understand what teaching entails?

The rest of the application letter is equally uninformative. Why is the student applying to this particular program? All Biology majors take Biology courses, and the student's grades are already on the transcript. What has the student learned from these courses that makes her want to pursue advanced study?

The same student could have written a much more effective letter by thinking about what admissions committees might be looking for and by documenting her strengths. Here is an example of the way this student might have rewritten her letter[1].

> To the admissions committee:
>
> Please consider my application for admission to your Ph.D. program in Biology. I will be graduating from Cerebral University in May with a B.S. in Botany. I believe that I have the experience and motivation to make a contribution to your program.
>
> I became interested in plant physiology through a seminar course taught by Professor Mendel. This was my first experience reading the original scientific literature, but by the end of the semester I was able to present a well-received research proposal on the subject of root growth, based largely upon my own library research.

---

[1] The research described in her letter is based on a paper by Kapulnik *et al.*, 1985. *Canadian Journal of Botany* 63: 627–631.

Dr. Mendel later invited me to conduct, in his laboratory, a research project examining the influence of nitrogen-fixing bacteria (<u>Azospirillum</u> spp.) on wheat seedling (<u>Triticum</u> <u>aestivum</u>) root development. Briefly, we first surface-treated wheat seeds in three percent sodium hypochlorite, to kill adhering microorganisms, and then incubated the seeds on moist filter paper in sterile, covered petri dishes. After the seeds germinated, some of the plates were inoculated with nitrogen-fixing bacteria and others were left untreated as controls.

After the seedlings had incubated for seven days at 25°C, I measured maximum seedling length (at 12X using a dissecting microscope fitted with a calibrated ocular micrometer), average seedling live weight (using a Cahn electrobalance precise to 0.1 μg), and average effective surface area of the root system (based on rate of HCl uptake, using the titration method of Carley and Watson, 1966). The presence of nitrogen-fixing bacteria significantly enhanced root development ($P < 0.05$, t-test) and shoot development ($P < 0.05$, t-test). We still do not know whether increased plant growth resulted directly from the presence of the bacteria (possibly through bacterial secretion of growth-stimulating chemicals); the root may be stimulated by the bacteria to release a growth-promoting hormone of its own.

Through this study, I learned the importance of careful experimental design and, more important, that I have the patience to do research. I do not wish to commit myself to a specific field of research at this time, believing that I would benefit from an additional year of literature and laboratory exploration, but I believe that I would like to study the role of hormones in early plant development.

During my last semester at Cerebral University, I have been acting as undergraduate teaching assistant for the introductory Biology laboratory. I find that having to explain things to other students forces me to come to grips with what I do and do not know. I am enjoying the challenge greatly, and I look forward to doing additional teaching in the future; I'm learning a lot about Biology through teaching.

I have asked the following faculty members at Cerebral University for letters of recommendation:

Professor G. Mendel (Biology Dept.)

Professor John Pilger (Biology Dept.)

Professor Marcia Stubbs (English Dept.)

Thank you for considering my application. I look forward to receiving your response.

Sincerely,

Dineen Arms

Dineen Arms

This letter, too, conveys enthusiasm, but it is an enthusiasm that reflects knowledge, experience, and commitment; the applicant seems to understand what research is all about and apparently knows how to go about doing it. Moreover, we see that Ms. Arms thinks clearly and writes well.

Dineen Arms would write a somewhat different letter if she were applying for a job as a technician in a research laboratory. In this letter, she would want to emphasize her skills and reliability as a laboratory worker and her interest in the type of research being done in the laboratory to which she is applying. An example of such a letter follows.

Dear Professor Bilderback:

Please consider my application for the technical position you advertised recently in the <u>Boston Globe</u>. I will be graduating from Cerebral University this May with a B.S. degree in Botany. Although I eventually expect to return to school to pursue a Ph.D., I would first like to work in a plant physiology laboratory for one or two years to learn some additional techniques and to become more familiar with various research fields and approaches.

I first became interested in plant physiology through a seminar course with Professor G. Mendel at Cerebral University. During the semester, we read two of your recent papers describing studies on the hormonal control of <u>Selaginella kraussiana</u> orientation to light.

Following this seminar, I began doing research on wheat seedlings (<u>Triticum aestivum</u>) in Professor Mendel's laboratory, investigating the influence of nitrogen-fixing bacteria (<u>Azospirillum</u> spp.) on root development. We found that root development was significantly enhanced by the presence of nitrogen-fixing bacteria. The next step in the study will be to determine whether increased root growth results directly from the presence of the bacteria (possibly through bacterial secretion of growth-stimulating chemicals), or indirectly, by stimulation of the root to secrete a growth-promoting hormone.

Through this study, I learned a variety of general laboratory techniques (use of assorted balances, sterile culture methodology, measurement of effective root surface area using the tritration technique of Carley and Watson, 1966), and I also learned that I have the patience and motivation needed to do careful research. I have taken five laboratory courses in Biology (General Genetics, Invertebrate Zoology, Comparative Animal Physiology, Cell Biology, and Developmental Biology), in which I learned several specialized laboratory techniques, including the pouring and use of electrophoretic gels, measurement of organismal and mitochondrial respiration rates, and use of an osmometer.

In short, I am very much interested in your research and believe I can make a contribution to the work of your laboratory. I have requested letters of recommendation from the following faculty at Cerebral University:

Professor G. Mendel (Biology Dept.)
Professor John Pilger (Biology Dept.)
Professor Marcia Stubbs (English Dept.)

Thank you for considering my application; I look forward to hearing from you.

Sincerely,

Dineen Arms

Dineen Arms

Both of these letters convey knowledge of the position or program applied for, a sincere interest in Biology, and a high level of ability and commitment. Your letter should do the same. Your credentials may not be as impressive as Ms. Arms's, but if you think about the experience you have had in relationship to the skills required for the position or program, you should be able to construct an effective letter. Take your letter through several drafts until you get it right (see Chapter 8, on revising).

RECRUITING EFFECTIVE LETTERS OF RECOMMENDATION

Letters of recommendation can be extremely important in determining the fate of your application. Although you do not write these letters yourself and rarely even get the opportunity to read them, you can take steps to increase their effectiveness.

Getting an *A* in a course does not guarantee a strong letter of recommendation from the instructor of that course. The most useful letters to admissions committees and prospective employers are those commenting on such characteristics as the following: laboratory skills, communication skills (written and oral), motivation, ability to use time efficiently, curiosity, maturity, intelligence, ability to work independently, and ability to work with others. Instructors cannot comment on these attributes unless you become more than a grade in their record books. Make an appointment to talk with some of your instructors about your interests and plans. We faculty members are usually happy for the opportunity to get to know students better.

When it is time to request letters of recommendation, choose three or four instructors who know something about your abilities and goals, and ask each of them if he or she would be able to support your application by writing a letter of recommendation. Give each person the opportunity to decline your invitation. If the people you ask agree to write on your behalf, make their task easier by giving them a copy of your résumé, transcript, and letter of application and, if appropriate, a copy of the job advertisement. Be certain to indicate clearly the application deadline and the address to which the recommendation should be sent.

It takes as much time and thought to write an effective letter of recommendation as it takes to write an effective letter of application. Don't lose good will by requesting letters at the last minute. Give your instructors at least two weeks to work on these letters. "It has to be in by this Friday" will probably annoy your prospective advocate and may not allow the recommender the time needed to prepare a good letter even if he or she is still in a cooperative mood. Moreover, last-minute requests don't speak favorably about your planning and organizing abilities, and they imply a lack of respect for your instructor. So be considerate, and thereby get the best recommendation possible.

8
Revising

The preceding chapters concern the reading, notetaking, thinking, synthesizing, and organizing that permit you to capture your thoughts and your evidence in a first draft. This chapter concerns the revising that must follow, in which you examine the first draft critically and diagnose and treat the patient as necessary.

Writing a first draft gives you the opportunity to get facts, ideas, and phrasings on paper, where they won't escape. Once you have captured your thoughts, you can concentrate on reorganizing and rephrasing them in the clearest, most logical way. It is difficult to revise your own work effectively unless you can examine it with a fresh eye. For this reason, try to complete your first draft at least several days before the final product is due. Plan ahead; be sure to allow time for careful revision. Reading your paper aloud — and listening to yourself as you read — often reveals weaknesses that you would otherwise miss. It also helps to have one or more fellow students carefully read and comment on your draft at this stage of its development; it is always easier to identify writing problems — wordiness, ambiguity, faulty logic, faulty organization, spelling and grammatical errors — in the work of others. Choose any system that feels comfortable, but always revise your papers before submitting them.

All writing benefits from revision. No matter how sound, or even brilliant, your thoughts and arguments are, it is the manner in which you express them that will determine whether or not they are understood and appreciated by readers. With pencil, scissors, and tape at the ready, the time has come to edit your first draft: for content, for clarity, for conciseness, for flow (coherence), and for spelling and grammar. If you are writing with a word processor, make your revisions on printed copy rather than on-screen; to edit effectively you must see more than one screen's-worth of text at a time. Continue editing and revising until your work is ready for the eyes of the instructor, admissions committee, or potential employer. This chapter should help you know when you have arrived at that point.

REVISING FOR CONTENT

Make sure every sentence says something. Consider the following opening sentence for an essay on the tolerance of estuarine fish to changes in salinity:

```
Salinity is a very important factor in marine

environments.
```

What does this sentence say? Is the author really trying to tell us that the ocean is salty? A careful editor will delete the sentence and begin anew with a sentence that says something worth reading. For example,

```
Estuarine fish may be subjected to enormous changes in

salinity within only a few hours.
```

The author of the revised opening sentence knows where his or her essay is headed, and so does the reader.

While revising for content, keep in mind an audience of your peers, not your instructor. In particular, be sure to define all scientific terms and abbreviations; it is not enough simply to use them properly. Brief definitions will help keep the attention of readers who may not know or may not remember the meaning of some terms and will also demonstrate to your instructor that you know the meaning of the specialized terminology you are using. Try to make your writing self-sufficient; the reader should not have to consult textbooks or other sources in order to understand what you are saying. As always, if you write so that you will understand your work years in the future, or so that your classmates will understand the work now, your papers and reports will generally have greater impact and will usually earn a higher grade.

REVISING FOR CLARITY

Be sure each sentence says what it's supposed to say; you want the reader's head to be nodding up and down, not side to side. Which way is the reader's head going in the following example?

```
These methods have different resorption rates and tail
shapes.
```

Do methods have tails? Can methods be resorbed? This sentence fails to communicate what its author had in mind. Indeed, it is difficult to tell *what* the author had in mind. Here is another sentence that does not reflect the intentions of its author:

```
From observations made in aquaria, feeding rates of the
fish were highest at night.
```

How many observers do you suppose can fit into an aquarium? Aquaria usually contain fish, not authors; is the author of our example all wet? A revised sentence might read:

```
Feeding rates of fish held in aquaria were highest at
night.
```

Walking buildings apparently are a problem on some college campuses:

```
With quick strides, the Science Center came into view.
```

Perhaps we should have the author doing the walking?

> With quick strides, I soon arrived at the Science Center.

Here is another sentence that simply is not doing its author's bidding:

> In order to keep the size of the samples constant, the
> sampling pipet was calibrated so that the volume of a
> single drop was known.

It is difficult to see how calibrating a pipet will keep sample sizes constant. Presumably, the author means that the same pipet was used for each sample (which would keep sample sizes constant). That the size of each sample was known is really a separate, independent thought.

Here are two additional examples of unclear scientific writing:

1. This determination was based on mannitol's relative
   toxicity to sodium chloride.
2. Swimming in fish has been carefully studied for only a
   few species.

How can one chemical be toxic to another chemical? I suspect the author is trying to tell us that two chemicals differ in their toxicity to some organisms. As for the second example, have you ever gone swimming in a fish, or even thought of attempting such a thing? No wonder it has been so little studied! We know what the author means, but he or she has not said it.

Confusing sentences surround us in our everyday worlds, too. Consider this example taken from the local newspaper:

> Offer void where prohibited by law, or while supplies
> last.

The meaning of this sentence is not immediately clear. It is apparently impossible for anyone to take advantage of this offer; the offer is either prohibited by law or, if permitted by law, is void while supplies last. Supplies can never run out, since the advertiser apparently is unwilling to fill your order as long as they have anything in stock. If supplies ever did run out, perhaps by eventual disintegration of the product, the advertiser could then honor your request; but they would no longer have anything to send you.

Make each sentence state its case unambiguously. Here is a sentence that does not:

```
Sea stars prey on a wide range of intertidal animals, de-
pending on their size.
```

Is the author talking about the size of the sea stars that are preying or about the size of the intertidal animals that are preyed upon? Don't be embarrassed at finding sentences like this one in early drafts of your papers and reports. Be embarrassed only when you don't edit them out of your final draft.

Frequent use of the pronouns *it*, *they*, and *them* in your writing should sound an alarm: Probable ambiguity ahead. Consider the following example of the troubles these words can cause:

```
The body is covered by a cuticle, but it is unwaxed.
```

Which is unwaxed: the body, or the cuticle? Similarly, *it* makes the second of the following two sentences equally ambiguous:

```
Ampullar muscles contract, pressurizing the chamber and
forcing water into the tube foot. It then elongates and
comes into contact with the substratum.
```

Is the tube foot elongating here, or is it the chamber? Here is an example where *they* creates a similar problem:

```
Tropical countries are home to both venomous and nonven-
omous snakes. They kill their prey by constriction or by
biting and swallowing them.
```

How much clearer the last sentence could become by replacing *they* with a few words of substance and by deleting *them* entirely:

```
Tropical countries are home to both venomous and nonven-
omous snakes. The nonvenomous snakes kill their prey by
constriction or by biting and swallowing.
```

One final example will show just how troublesome *they* can be:

```
When the larval stage of the parasitic worm was exposed
only to animals of a species that never serves as a host,
they did not parasitize them.
```

If *they* have their way, the reader must guess who is not parasitizing whom. Realizing that the sentence is in difficulty, we revise:

```
When the larval stage of the parasitic worm was exposed
to test animals of a species that never serves as a host,
the larvae did not parasitize the test animals.
```

In short, when editing your work you must read carefully and with scepticism, checking to be sure that you have said exactly what you mean. Never make the reader guess what you have in mind. Never give the reader cause to wonder whether, in fact, you have anything in mind. Everything you write must make sense — to yourself and to the reader. As you read each sentence you have written, think: What does this sentence say? What did I mean it to say? Make each sentence work on your behalf, leading the reader easily from fact to fact, from thought to thought.

Please note that you need not be a grammarian to write correctly and clearly. With a little practice, especially if you read your work aloud, you can quickly learn to recognize a sentence in difficulty and sense how to fix it without ever knowing the name of the grammatical rule that has been violated.

REVISING FOR COMPLETENESS

Make sure each thought is complete. Be specific in making assertions. The following statement is much too vague:

Many insect species have been described.

How many is "many"? After editing, the sentence might read,

Nearly one million insect species have been described.

Similarly, the following sentence

More caterpillars chose diet <u>A</u> than diet <u>B</u> when given a
choice of the two diets (Fig. 2).

would benefit from this alteration:

Nearly five times as many caterpillars chose diet <u>A</u> than
diet <u>B</u> when given a choice of the two diets (Fig. 2).

Be especially careful to revise for completeness whenever you find that you have written *etc.*, an abbreviation for the Latin term *et cetera*, meaning "and others" or "and so forth." In writing a first draft, use *etc.*'s freely when you'd rather not interrupt the flow of your thoughts by thinking about exactly what "others" you have in mind. When revising, however, replace each *etc.* with words of substance; in scientific writing, an *etc.* makes the reader suspect fuzzy thinking. You should find yourself thinking, "What, exactly, *do* I have in mind here?" If you come up with additional items for your list, add them. If you find that you have nothing to add, simply replace the *etc.* with a period and you will have produced a shorter, clearer, sentence. Consider the following sentence and its two improvements:

ORIGINAL VERSION

```
Plant growth is influenced by a variety of environmental
factors, such as light intensity, nutrient availability,
etc.
```

REVISION 1

```
Plant growth is influenced by a variety of environmental
factors, such as light intensity, day length, nutrient
availability, and temperature.
```

REVISION 2

```
Plant growth is influenced by such environmental factors
as light intensity, day length, and nutrient
availability.
```

In the original version, the author has dodged the responsibility of clear writing, forcing the reader to determine what is meant by *etc.* The sentence, although grammatically correct, is incomplete, waiting for the reader to fill in the missing information. The reader may justifiably wonder whether the writer knows what other factors affect plant growth. Both revised versions clearly indicate what the author had in mind. Revising for completeness often requires you to return to your notes or to the sources upon which your notes are based.

REVISING FOR CONCISENESS

Omitting unnecessary words will make your thoughts clearer and more convincing. In particular, such phrases as, "It should be noted that," "It is interesting to note that," and "The fact of the matter is that" commonly creep into first drafts, but should be ruthlessly eliminated in preparing the second. Such verbal excess also takes less conspicuous forms. How might you shorten this next sentence?

```
Dr. Smith's research investigated the effect of pesti-
cides on the reproductive biology of birds.
```

Who did the work: Dr. Smith or his research? A reasonable revision would be:

```
Dr. Smith investigated the effect of pesticides on the
reproductive biology of birds.
```

We have eliminated one word and the sentence has not suffered a bit. Working on the sentence further, we can replace "the reproductive biology of birds" with "avian reproduction," achieving a net reduction of three more words:

```
Dr. Smith investigated the effect of pesticides on avian
reproduction.
```

The next example requires more ruthless pruning:

> The results indicated a role of hemal tissue in moving
> nutritive substances to the gonads of the animal.

Any sentence containing such a long string of prepositional phrases — of tissue, in moving substances, to the gonads, of the animal — is automatically a candidate for the editor's operating table. This sentence actually contains a simple thought, buried amidst a clutter of unnecessary words. After surgery, the thought emerges clearly:

> The results indicated that hemal tissue moved nutrients
> to the animal's gonads.

Revising for conciseness, "Plant vascular tissues function in the transport of food through xylem and phloem" becomes, "Plant vascular tissues (xylem and phloem) transport food" and "Gould therefore arrives at the conclusion that . . ." becomes "Gould therefore concludes that . . . ." "Schooling of fish is a well documented phenomenon" becomes "Schooling of fish is well documented," or even "Fish schooling is well documented."

The passive voice is often a great enemy of concise writing. If the subject is on the receiving end of the action, the voice is passive:

> Rats and mice were experimented on by him.

If, on the other hand, the subject of a sentence is on the delivering end of the action, the voice is said to be active:

> He experimented with rats and mice.

Note that the "active" sentence contains only six words, while its "passive" counterpart contains eight. In addition to creating excessively wordy sentences, use of the passive voice often makes the active agent anonymous and a weaker, sometimes ambiguous sentence may result:

```
Once every month for two years, mussels were collected
from five intertidal sites in Barnstable County, MA.
```

Who should the reader contact if there is a question about where the mussels were collected? Were the mussels collected by the writer, by fellow students, by an instructor, or by a private company? Eliminating the passive voice clarifies the procedure:

```
Once every month for two years, members of the class col-
lected mussels from five intertidal sites in Barnstable
County, MA.
```

Similarly, "It was found that" becomes "I found," or "we found," or, perhaps, "Smith (1986) found." Whenever it is important, or at least useful, that the reader know who the agent of the action is, and whenever the passive voice makes a sentence unnecessarily wordy, use the active voice:

Passive: Little is known of the nutritional requirements of these animals.

Active: We know little about the nutritional requirements of these animals.

Passive: The results were interpreted as indicative of . . .

Active: The results indicated . . .

Passive: In the present study, the food value of seven diets was compared, and the chemical composition of each diet was correlated with its nutritional value.

Active: In this study, I compared the food value of seven diets and correlated the chemical composition of each diet with its nutritional value.

Note in this last example that it is perfectly acceptable to use the
pronoun I in scientific writing; switching to the active voice expresses
thoughts more forcibly and clearly and often eliminates unnecessary
words. Be a person of few words; your reader will be grateful.

REVISING FOR FLOW

A strong paragraph — indeed, a strong paper — takes the
reader smoothly and inevitably from a point upstream to one
downstream. Link your sentences and paragraphs using appropriate
transitions, so that the reader moves effortlessly and inevitably
from one thought to the next, logically and unambiguously. Min-
imize turbulence. Always remind the reader of what has come
before, and help the reader anticipate what is coming next. The
first of the two following examples gives the reader a choppy ride
indeed, and cries out for careful revision, not of the ideas themselves
but of the way they are presented. A suitable revision, which
facilitates the flow of ideas, is presented in the second example.
Note especially the use of two important transitional words and
phrases, *thus* and *in contrast to*.

Since aquatic organisms are in no danger of drying out,
gas exchange can occur across the general body surface.
The body walls of aquatic invertebrates are generally
thin and water permeable. Terrestrial species that rely
on simple diffusion of gases through unspecialized body
surfaces must have some means of maintaining a moist body
surface, or must have an impermeable outer body surface
to prevent dehydration; gas exchange must occur through
specialized, internal respiratory structures.

Since aquatic organisms are in no danger of drying out,

```
gas exchange can occur across the general body surface.
Thus, the body walls of aquatic invertebrates are gener-
ally thin and water-permeable, facilitating such gas ex-
change. In contrast to the simplicity of gas exchange
mechanisms among aquatic species, terrestrial species
that rely on simple diffusion of gases through unspe-
cialized body surfaces must either have some means of
maintaining a moist outer body surface, or must have an
impermeable outer body covering that prevents dehydra-
tion. If the outer body wall is impermeable to water and
gases, respiratory structures must be specialized and
internal.
```

In the first draft, the reader must struggle to find the connection between sentences. In the revised version, the writer has assisted the reader by connecting the thoughts, resulting in a more coherent paragraph.

Here is one more example of a stagnating paragraph that carries its reader nowhere:

```
The energy needs of a resting sea otter are three times
those of terrestrial animals of comparable size. The sea
otter must eat about 25% of its body weight daily. Sea
otters feed at night as well as during the day.
```

Revising for improved flow, or coherence, produces the following paragraph. Note that the writer has introduced no new ideas. The additions, here underlined, are simply clarifications that make the connections between each point explicit.

```
The energy needs of a resting sea otter are three times
```

those of terrestrial animals of comparable size. <u>To sup-</u>
<u>port such a high metabolic rate</u>, the sea otter must eat
about 25% of its body weight daily. <u>Moreover</u>, sea otters
feed continually, at night as well as during the day.

The following transitional words and phrases are especially
useful in linking thoughts to improve flow: *in contrast, however,*
although, thus, whereas, even so, nevertheless, moreover, despite the, in
addition to. Repetition is also a commonly used and effective way
to link thoughts. For instance, repetition has been used to connect
the first two sentences of the revised example about sea otters:
"To support such a high metabolic rate" essentially repeats, in
summary form, the information content of the first sentence. Repetition is a particularly effective way of linking paragraphs; in
reminding the reader of what has come before, the author con-
solidates his or her position and then moves on. Use these and
similar transitions to move the reader smoothly from the beginning
of your paper to the end.

Judicious use of the semicolon can also ease the reader's
journey. In particular, a semicolon is appropriate when the second
sentence of a pair explains or clarifies something contained in the
first:

This enlarged and modified bone, with its associated
muscles, serves as a useful adaptation for the panda.
With its ''thumb,'' the panda can easily strip the bamboo
on which it feeds.

With this construction, the reader would probably have to pause
to consider the connection between the two sentences. Using a
semicolon, the passage would read:

This enlarged and modified bone, with its associated
muscles, serves as a useful adaptation for the panda;

```
with its ''thumb,'' the panda can easily strip the bamboo
on which it feeds.
```

The semicolon links the two sentences and eliminates an obstruction in the reader's path. Similarly, a semicolon provides an effective connection between thoughts in the following two examples:

```
Recently, we demonstrated the rapid germination of rad-
ish seeds; nearly 80% of the seeds germinated within
three days of planting.
```

```
Ciliated protozoans, such as Paramecium aurelia, repro-
duce primarily by the asexual process of binary fission;
a single individual can quickly give rise to a large
population.
```

REVISING FOR TELEOLOGY

Remember, animals do not act or evolve with intent (p. 9). Consider the following examples of teleological writing and learn to recognize the trend in your own work:

```
Barnacles are incapable of moving from place to place,
and therefore had to evolve a specialized food-collect-
ing apparatus in order to survive.
```

```
Aggression is a directed behavior that many sea anemones
exhibit to promote the survival of an individual's own
genotype.
```

Revise all teleology out of your writing.

REVISING FOR SPELLING ERRORS

If you would not be willing to bet your most prized possession that you have correctly spelled a particular word, look up the word in a dictionary. Misspellings convey the impression of care-lessness, laziness, or perhaps even stupidity. These are not advisable images to present to instructors, prospective employers, or the admissions officers of graduate or professional programs.

It helps to keep a list of words that you find yourself using often and consistently misspelling. *Desiccation* was on my list for quite some time; *proceed* and *precede* are still on it. If you are uncertain of the spelling of a particular word and absolutely cannot bear the thought of opening the dictionary yet again, change the word to a synonym that you can spell.

A few peculiarities of the English language are worth pointing out:

1. *Mucus* is a noun; as an adjective, the same slime becomes *mucous*. Thus, many marine animals produce mucus, and mucous trails are produced by many marine animals.
2. *Seawater* is always a single word. *Fresh water*, however, is two words as a noun and one word as an adjective. Thus, freshwater animals live in fresh water.

And don't forget to underline species names: <u>Littorina littorea</u> (the periwinkle snail), <u>Chrysemys picta</u> (the eastern painted turtle), <u>Taraxacum officinale</u> (the common dandelion), <u>Homo sapiens</u> (the only animal that writes laboratory reports).

REVISING FOR GRAMMAR AND PROPER WORD USAGE

Appendix C (p. 182) lists a number of books that include excellent sections on grammar and proper word usage. While on the lookout for sentence fragments, run-on sentences, faulty use of commas, faulty parallelism, incorrect agreement between subjects and verbs, and other grammatical blunders, you should also be on the lookout for violations of five especially troublesome rules of usage when revising your work:

1. *between/among. Between* (from *by twain*) usually refers to only two things:

```
The 20 caterpillars were randomly distributed between

the two dishes.
```

Among usually refers to more than two things:

```
The 20 caterpillars were randomly distributed among the

eight dishes.
```

2. *which/that.* Most of your *which*'s should be *that*'s.

```
A fish that lives at a depth of 1000 m is exposed to 100

atmospheres of pressure.
```

```
This fish, which lives at depths up to 1000 m, experi-

ences up to 100 atmospheres of pressure.
```

In the first example, *that* introduces a defining, or restrictive clause; we are being told about a particular fish, or type of fish. On the other hand, the *which* of the second example introduces a nondefining, or nonrestrictive clause. The introduced phrase is, in effect, an aside, adding extra information about the fish in question. The sentence would survive without it.

Improper use of *that* and *which* can occasionally lead to ambiguity or falsehood. Consider the following sentence about cephalopods, a group of molluscs including the squid and octopus:

```
In cephalopods that lack calcified shells, locomotion is

accomplished entirely by contractions of the muscular

mantle.
```

This sentence refers only to one group of cephalopods, those that lack shells. Again, *that* is introducing a restrictive clause. Replacing *that* with *which* drastically changes the meaning of the sentence:

```
In cephalopods, which lack calcified shells, locomotion
is accomplished entirely by . . .
```

Here the reader is told, in a non–defining clause, that all cephalopods lack calcified shells; this is simply not true.

As in the examples given, *which* is commonly preceded by a comma. When deciding between *which* and *that*, if the word doesn't need a comma before it for the sentence to make sense, the right word is probably *that*.

3. *its/it's*. *It's* is always an abbreviated form of *it is*. If *it is* does not belong in your sentence, use the possessive pronoun *its*.

```
When treated with the chemical, the protozoan lost its
cilia and died.

It's clear that the loss of cilia was caused by treatment
with the chemical.
```

While we're at it, let's revise that last sentence to eliminate the passive voice:

```
It's clear that treatment with the chemical caused the
loss of cilia.
```

In general, contractions are not welcome in formal scientific writing. Thus, you can avoid the problem entirely by writing *It is* when appropriate:

```
It is clear that treatment with the chemical caused the
loss of cilia.
```

4. *effect/affect*. *Effect* as a noun refers to a result or outcome:

```
What is the effect of fuel oil on the feeding behavior of
sea birds?
```

Effect as a verb means *to bring about*:

```
What changes in feeding behavior will fuel oil effect in
sea birds?
```

Affect as a verb means *to influence* or *to produce an effect upon*:

```
How will the fuel oil affect the feeding behavior of sea
birds?
```

5. *i.e./e.g.* Students commonly use these two abbreviations interchangeably. Nevertheless, *i.e.* is an abbreviation for *id est*, which in Latin, means *that is* or *that is to say*. For example,

```
Data on sex determination suggest that this species
has only two sexual genotypes, i.e., female (XX) and male
(XY).
```

```
The embryos were undifferentiated at this stage of
development; i.e., they lacked external cilia and the
gut had not yet formed.
```

In contrast, *e.g.* stands for *exempli gratia*: *for example*. I will give two examples of its use:

```
During the precompetent period of development, the
larvae cannot be induced to metamorphose (e.g., Crisp,
1974; Bonar, 1978; Chia, 1978; Hadfield, 1978).
```

> However, the larvae of several butterfly species
> (e.g., <u>Papilio</u> <u>demodocus</u> Esper, <u>P</u>. <u>eurymedon</u>, and <u>Pieris</u>
> <u>napi</u>) are able to feed and grow on plants that the adults
> never lay eggs on.

In the first case, the writer uses *e.g.* to indicate that what follows is only a partial listing of references supporting the statement: "for example, see these references," in other words. In the second case, the writer indicates that he gives only a partial list of butterfly species that don't lay eggs on all suitable plants.

6. And don't forget: *The data are* . . . (see page 9).

BECOMING A GOOD REVISOR

The best way to become an effective revisor of your own writing is to become a critical reader of other people's writing. Whenever you read a newspaper, magazine, or textbook, be on the lookout for ambiguity and wordiness, and think about how the sentence or paragraph might best be rewritten. You will gradually come to recognize the same problems, and the solutions to these problems, in your own writing. As a first step in the right direction, read the following sentences and try to verbalize the ailment afflicting each one. Then revise those sentences that need help. Pencil your suggested changes directly onto the sentences, using the guide presented in Table 5 (p. 169) and the following example.

EXAMPLE

Hermaphroditism is ¢ommonly encountered among invertebrates. For example, the young East Coast oyster, <u>Crassostrea</u> <u>virginica</u>, matures as a male, later ƒecomes a female and may change sex every few years there after. sequential hermaphrodites generally change sex only once, and usually change from male to female. In contrast to species that change sex as they age, many invertebrates are simultaneous hermaphodites. Self-fertilization is rare among simultaneous hermaphrodites, although it can occur, as in the tapeworms.

Table 5.   Proofreader's Symbols Used in Revising Copy

Problem	Symbol	Example
1. Word has been omitted	∧ (caret)	study describes *the* effect ∧
2. Letter has been omitted	∧ (caret)	that bo*o*k ∧
3. Letters are transposed	∿	f*ro*m the sea
4. Words are transposed	∿	was (only exposed)
5. Word should be capitalized	≡ (three short underlines)	these data ≡
6. Word should be lower-case	/ (slash)	/hese Data
7. Word should be in italics	___ (underline once)	Homo sapiens
8. Words are run together		(draw vertical line between)	edit|carefully
9. Word should be deleted	⎯ (draw line through)	the ~~nice~~ data
10. Space should not have been left	◡ (sideways parentheses)	the e◡nd
11. Wrong letter	/ (draw line through and add correct letter above)	*f*/emale
12. Wrong word	⎯ (draw line through and add correct word above)	*These* ~~This~~ data
13. Need to begin a new paragraph	¶ (paragraph symbol)	female. ¶ In contrast
14. Restore original	(STET)	The ~~energy~~ needs (STET)

It is wise when editing someone else's work to use a different color pen or pencil to be sure the reader will see the suggested change.

1. To perform this experiment there had to be a low tide. We conducted the study at Blissful Beach on September 23, 1986, at 2:30 PM.

2. In *Chlamydomonas reinhardi*, a single-celled green alga, there are two matine types, + and −. The + and − cells mate with each other when starved of nitrogen and form a zygote.

3. Protruding form this carapace is the head, bearing a large pair of second antennae.

4. The order in which we think of things to write down is rarely the order we use when we explain what we did to a reader.

5. The purpose of Professor Wilson's book is the examination of questions of evolutionary significance.

6. This data is summarized in Table 1.

7. One example of this capcity is observed in the phenomenon of encystment exhibited by many fresh water and parasitic species.

8. In a sense, then, the typical protozoan may be regarded as being a single-celled organism.

9. An estuary is a body of water nearly surrounded by land whose salinity is influenced by freshwater drainage.

10. The résumé presents a summary of your educational background, research experience and goals.

11. In textbooks and many lectures, you are being presented with facts and interpretations.

12. What is it that bothers us about the egg and sperm of that couple who died in Australia sitting on ice?

13. It should be noted that analyses were done to determine whether the caterpillars chose the different diets at random.

14. These experiments were conducted to test whether the condition of the biological films on the substratum surface triggered settlement of the larvae.

15. Various species of sea anemones live throughout the world.

16. This data clearly demonstrates that growth rates vary with temperature.

17. Hibernating mammals mate early in the spring so that their

offspring can reach adulthood before the beginning of the next winter.

18. This study pertains to the investigation of the effect of this pesticide on the orientation behavior of honey bees.

19. The results reported here have led the author to the conclusion that thirsty flies will show a positive response to all solutions, regardless of sugar concentration (see figure 2).

20. Numbers are difficult for listeners to keep track of when they are floating around in the air.

21. Those seedlings possessing a quickly growing phenotype will be selected for, whereas . . .

22. Under a dissecting microscope, a slide with a drop of the culture was examined at 50X.

23. Measurements of respiration by the salamanders typically took one-half hour each.

There are several ways to improve each of the preceding sentences. For reference, my revisions are shown in Appendices D and E (pp. 184–190), but you should make your own modifications before looking at mine. Be sure that you can identify the problem suffered by each original sentence, that you understand how that problem was solved by my revision, and that your revision also solves the problem (and does not introduce any new difficulties).

9
Answering Essay Questions

Answering essay questions on examinations differs from the other forms of scientific writing already discussed only in two respects: The essay examination must be completed within a short time, usually from 15 to 50 minutes, and you no longer have a choice in the subject of the essay. A winning answer to an essay question will follow all the guidelines outlined in Chapter 1. Your performance on essay questions can be strengthened by keeping in mind a few additional points:

1. Read the question carefully before writing anything. You must answer the question posed, not the question you would have preferred to see on the examination. In particular, note whether the question asks you to list, discuss, or compare. A list will not satisfy the requirements of a discussion or comparison. A request for a list tests to see whether you know all components of the answer; a request for a discussion additionally examines your understanding of the interrelationships among these components.

Consider this list of the characteristics of a Big Mac® and a Whopper®, based on a study conducted in Somerville, Massachusetts, in May of 1985:

Big Mac	*Whopper*
2 beef patties	1 beef patty
patties 3.25″ diameter	patty 3.75–4″ diameter
fried beef	broiled beef

3-part bun (3 slices)	2-part bun (top & bottom)
sesame seeds on top bun	sesame seeds on top bun
slice of pickle	slice of pickle
chopped onion	slices of onion
slice of cheese	2–3 slices of tomato
lettuce	ketchup
sauce	mayonnaise
$1.54	$1.45
packed in styrofoam box	packed in cardboard box

Suppose you are asked to write an essay presenting the features of both items. Your essay might look like this:

> The Big Mac consists of two patties of fried ground beef, each patty approximately 3.25 inches in diameter, with lettuce, chopped onion, sliced pickle, a slice of cheese, some reddish sauce, and a three-part bun, with the two patties separated from each other by one of the slices of bun. The top slice of the bun is covered with sesame seeds. The Big Mac sells for $1.54 and is served in a styrofoam box.
>
> The Whopper consists of one slice of broiled ground beef (approximately four inches in diameter), with mayonnaise, ketchup, several slices each of tomato, pickle, and onion, and a two-part bun, with the upper half of the bun covered with sesame seeds. The Whopper sells for $1.45 and is served in a cardboard box.

If you are asked to compare, or to compare and contrast, the two products, your essay must be written differently:

> Both the Big Mac and the Whopper contain ground beef and are served on buns. The two hamburgers differ, how-

ever, with regard to the way the meat is cooked, the way
the meat and bread are distributed within the hamburger,
the nature of accompanying condiments, and the price.

The meat in the Big Mac is fried, and each sandwich
contains two patties, each approximately 3.25 inches in
diameter and separated from the second patty by a slice
of bun. In contrast, the meat in the Whopper is broiled,
and each sandwich contains a single, larger patty, approximately 3.75—4 inches in diameter. The top bun of
both sandwiches is dotted with sesame seeds. Both the Big
Mac and the Whopper contain lettuce, onion, and slices of
pickle. The Big Mac, however, contains chopped onion,
whereas the onion in the Whopper is sliced. Moreover, the
Big Mac has a slice of cheese, which is absent from the
Whopper. On the other hand, the Whopper comes with slices
of tomato, which are absent from the Big Mac. Both sandwiches contain a sauce: ketchup and mayonnaise in the
Whopper and a pre-mixed sauce in the Big Mac. In comparison with the Big Mac, the Whopper is slightly less expensive ($1.45 versus $1.54).

If you are asked for a comparison and respond with a list,
you will probably lose points, not because your instructor is being
picky but because you have failed to demonstrate your understanding
of the relationship between the characteristics of the two products.
It is not the instructor's job to guess at what you understand; it
is your job to demonstrate what you know to the instructor. Note
that the facts included are the same in the two essays. The difference
lies in the way the facts are presented.

When asked for a list, give a list; this response requires less
time than a discussion, giving you more time to complete the rest
of the examination. When asked for a discussion, discuss: Present

the facts and support them with specific examples. When asked for a comparison, you will generally discuss similarities and differences, but the word *compare* can also mean that you should consider only similarities. Often an instructor will ask you to compare and contrast, avoiding any such ambiguity. If you have any doubts about what is required, ask your instructor during the examination.

   2. Present all relevant facts. Although there are many ways to answer an essay question correctly, your instructor will undoubtedly have in mind a series of facts that he or she would like to see included in your essay. That is, the ideal answer to a particular question will contain a finite number of components; the way you deal with each of these components is up to you, but each of the components should be considered in your answer.

   Before you begin to write your essay, then, list all components of the ideal answer, drawing both from lecture material and from any readings you were assigned. For example, suppose you are asked the following question:

> Discuss the influence of physical and biological factors on the distribution of plants in a forest.

What components will the perfect answer to this question contain? Begin by making a list of all relevant factors as they occur to you — don't worry about the order in which you jot these factors down:

Physical	*Biological*
amount of rainfall	competition with other plants
annual temperature range	
light intensity	predation by herbivores
hours of light per day	
type of soil	
pesticide use	
nutrient availability	

This list is not your answer to the essay question; it is an organizing vehicle intended for your use alone. Feel free to abbreviate, especially if pressed for time ("nutr. avail.," "pred. by herbs"), but be certain you won't misunderstand your own notes while writing the essay.

In preparing to write your answer to the essay question, arrange the elements of your list in some logical order, perhaps from most to least important or so that related elements are considered together; this grouping and ordering is most quickly done by simply numbering the items in your list in the order that you decide to consider them. You have now outlined your answer; the most difficult part of the ordeal is finished.

Incorporate into your essay each of the ordered components in your list. Avoid spending all your time discussing a few of these components to the exclusion of the others. If you discuss only four of the eight relevant issues, your instructor will be forced to assume you don't realize that the other issues are also relevant to the question posed. Show your instructor you know all the elements of a complete answer to the question.

3. Stick to the facts. An examination essay is not an exercise in creative writing and is not the place for you to express personal, unsubstantiated opinion. As with any other type of examination question, your instructor wishes to discover what you have learned and what you understand. Focus, therefore, on the facts and, as with all other forms of scientific writing, support all statements of fact or opinion with evidence or example. You may wish to suggest a hypothesis as part of your essay; if so, be sure to include the evidence upon which your hypothesis is based.

4. Keep the question in mind as you write. Don't include superfluous information. If what you write is irrelevant to the question posed, you probably won't get additional credit for your answer, and you will most likely annoy your instructor. If what you write is not only irrelevant but also wrong, you will probably lose points. By letting yourself wander off on tangents you will usually gain nothing, possibly lose points, and probably lose your instructor's good will; certainly, you will waste time that might more profitably be applied elsewhere on the examination. Listing the components of your answer before you write your essay will help keep you on track.

Appendix A
Means, Variances, Standard Deviations, and Standard Errors

Suppose you have two samples of three rats each. The rat tail lengths in Samples *A* and *B* are:

$$A = 7.0, 7.0, 7.0 \text{ cm}$$
$$B = 3.6, 14.1, 3.3 \text{ cm}$$

Both samples have the same mean value (7.0 cm), but *A* is much less variable than *B*. Simply listing the mean value, then, omits an important component of the story contained in your data.

The *variance* (σ^2) about the mean gives an indication of how variable your data are from one observation to the next. If you have access to a statistical calculator, just push the right buttons and you're almost done. If you are less fortunate, make your calculations using this formula:

$$\sigma^2 = \frac{\sum\limits_{i=1}^{N}(X_i - \overline{X})^2}{N-1}$$

N is the number of observations made, X_i is the value of the ith observation and $\overline{X}$ is the mean value of all the observations made in a sample.

Σ is the symbol for summation. In this case, you are to sum the squared differences of each individual measurement from the mean of all the measurements. As an example, suppose you have the following data points:

$$5 \text{ cm}$$
$$4$$
$$4 \qquad\qquad N = 5$$
$$6$$
$$5$$

$$\overline{X} = \frac{\sum\limits_{i=1}^{N}}{N} = \frac{24}{5} = 4.8 \text{ cm}$$

$$\sigma^2 = \frac{(5 - 4.8)^2 + (4 - 4.8)^2 + (4 - 4.8)^2 + (6 - 4.8)^2 + (5 - 4.8)^2}{4}$$
$$= 0.7$$

All you are doing is seeing how far each observation is from the mean value obtained and adding all these variations together. The squaring is done simply to eliminate minus signs, so that you have only positive numbers to work with. Clearly, 100 measurements should give you a more accurate estimation of the true mean tail length than only ten measurements, and, if you had the time and the patience, 1,000 measurements would be better still. We thus divide the sum of the individual variations by a factor related to the number of observations made. Increasing the sample size will reduce the extent of experimental uncertainty. Variance, then, is a measure of the amount of confidence we can have in our measurements. The smallest possible variance is zero (all samples were identical); there is no limit to the potential size of the variance.

To calculate the standard deviation (SD), simply take the square root of the variance.

To calculate the standard error of the mean, simply divide the standard deviation by the square root of N.

Appendix B
Commonly Used Abbreviations

	Abbreviation	*Example*
Length:		
meter	m	3 m
centimeter (10^{-1} meter)	cm	15 cm
millimeter (10^{-3} meter)	mm	4.5 mm
micron (10^{-6} meter)	μm	5 μm
Weight:		
gram	g	10 g
kilogram (10^3 grams)	kg	15 kg
milligram (10^{-3} gram)	mg	16 mg
microgram (10^{-6} gram)	μg	4 μg
Volume:		
liter	l	3 l
milliliter (10^{-3} liter)	ml	37 ml
microliter (10^{-6} liter)	μl	13 μl

	Abbreviation	*Example*
Time:		
days	d	2 d
hours	h, hr	48 h, or 48 hr
seconds	s, sec	60 s, or 60 sec
Concentration:		
milliosmoles/liter	mOsm/l	650 mOsm/l
molar	M	a 0.3 M solution
salinity (parts per thousand)	‰ S, ppt	31‰ S seawater, or 31 ppt
parts per million	ppm	0.2 ppm copper
parts per billion	ppb	200 ppb copper
Statistics:		
mean	$\overline{X}$	$\overline{X}$ = 27.2 g/ individual
standard deviation	SD	SD = 0.8
standard error	SE	SE = 0.3
sample size	N	N = 16
Other:		
photoperiod (h light:h dark)	L:D	10L:14D
one species	sp.	*Crepidula* sp.
two or more species	spp.	*Crepidula* spp.
approximately	c., $\approx$	c. 25°C, or $\approx$25°C

Appendix C
Suggested References for Further Reading

GENERAL BOOKS ABOUT WRITING

Barnet, S., and M. Stubbs. 1986. *Practical Guide to Writing,* 5th ed. Boston: Little, Brown and Co.

Hall, D. 1985. *Writing Well,* 5th ed. Boston: Little, Brown and Co.

Maimon, E. P., G. L. Belcher, G. W. Hearn, B. F. Nodine, and F. W. O'Connor. 1981. *Writing in the Arts and Sciences.* Cambridge: Winthrop Publishing.

Strunk, W., Jr., and E. B. White. 1979. *The Elements of Style,* 3d ed. New York: The Macmillan Co.

BOOKS ABOUT SCIENTIFIC WRITING

Day, R. A. 1983. *How to Write and Publish a Scientific Paper,* 2d ed. Philadelphia: ISI Press.

King, L. S. 1978. *Why Not Say It Clearly? A Guide to Scientific Writing.* Boston: Little, Brown and Co.

Zinsser, W. 1985. *On Writing Well. An Informal Guide to Writing Nonfiction,* 3d ed. New York: Harper & Row.

TECHNICAL GUIDE FOR BIOLOGY WRITERS

CBE Style Manual Committee. 1983. *Council of Biology Editors Style Manual: A Guide for Authors, Editors, and Publishers in the Biological Sciences,* 5th ed. Washington, D.C.: Council of Biology Editors.

Appendix D
Revised Sample Sentences

1. ~~To perform this experiment there had to be a low tide.~~ We
conducted the study at Blissful Beach ^at low tide^ on September 23, 1984,
~~at 2:30 PM.~~

2. In <u>Chlamydomonas</u> <u>reinhardi</u>, a single-celled green alga, there
are two mating types, + and − . The + and − cells mate
with each other when starved of nitrogen, and form a zygote.

3. Protruding from this carapace is the head, bearing a large
pair of second antennae.

4. The order in which we think of things to write down is
rarely the order we use when ~~we~~ explain^ing^ what we did to a
reader.

5. ~~The purpose of~~ Professor Wilson's book ~~is the~~ examination*es*
~~of~~ questions of evolutionary significance.

184

6. These data are summarized in Table 1.

7. One example of this capacity is the encystment exhibited by many fresh water and parasitic species.

8. In a sense, then, the typical protozoan is a single-celled organism.

9. An estuary is a body of water nearly surrounded by land and whose salinity is influenced by freshwater drainage.

10. The résumé summarizes your educational background, research experience, and goals.

11. Textbooks and many lectures present you with facts and interpretations.

12. What is it that bothers us about the frozen egg and sperm of that couple who died in Australia?

13. The data were analyzed to determine whether the caterpillars chose the different diets at random.

14. These experiments test whether the con-

~~dition of the~~ biological films ~~on the substratum surface~~ triggered [surface] [larval]

settlement ~~of the larvae~~.

15. ~~Various species of sea anemones live throughout the world.~~
(*Delete sentence for lack of content.*)

16. Th~~is~~ [ese] data clearly demonstrate~~s~~ that growth rates vary with temperature.

17. Hibernating mammals mate early in the spring~~o so that~~ [As a consequence] their offspring ~~can~~ reach adulthood before the beginning of the next winter.

18. This study ~~pertains to the investigation of~~ [describes] the effect of this pesticide on the orientation behavior of honey bees.

19. ~~The results reported here have led the author to the conclusion~~ ~~that~~ thirsty flies ~~will~~ [apparently] show a positive response to all solutions, regardless of sugar concentration (~~see~~ figure 2).

20. (Numbers) are difficult for listeners to keep track of ~~when they are~~ (floating around in the air)

21. Those seedlings ~~possessing a quickly growing phenotype~~ [genetically programmed for faster growth] will be selected for, whereas . . .

22. ~~Under~~ using a dissecting microscope a slide with a drop of the culture was examined at 50X.

23. Measurements of salamander respiration ~~by the salamanders~~ typically took one-half hour each.

Appendix E
The Revised Sample Sentences in Final Form

1. We conducted the study at Blissful Beach at low tide on September 23, 1984.

2. In <u>Chlamydomonas reinhardi</u>, a single-celled green alga, there are two mating types, + and −. When starved of nitrogen, the + and − cells mate with each other and form a zygote.

3. Protruding from this carapace is the head, bearing a pair of large second antennae.

4. The order in which we think of things to write down is rarely the order we use when explaining to a reader what we did.

5. Professor Wilson's book examines questions of evolutionary significance.

6. These data are summarized in Table 1.

7. One example of this capacity is the encystment exhibited by many freshwater and parasitic species.

8. In a sense, then, the typical protozoan is a single-celled organism.

9. An estuary is a body of water nearly surrounded by land and whose salinity is influenced by freshwater drainage.

10. The résumé summarizes your educational background, research experience, and goals.

11. Textbooks and many lectures present you with facts and interpretations.

12. What is it that bothers us about the frozen egg and sperm of that couple who died in Australia?

13. The data were analyzed to determine whether the caterpillars chose the different diets at random.

14. These experiments tested whether the biological surface films triggered larval settlement.

15. (Sentence deleted for lack of content.)

16. These data clearly demonstrate that growth rates vary with temperature.

17. Hibernating mammals mate early in the spring. As a consequence, their offspring reach adulthood before the beginning of the next winter.

18. This study describes the effect of this pesticide on the orientation behavior of honey bees.

19. Thirsty flies apparently show a positive response to all solutions, regardless of sugar concentration (Fig. 2).

20. Numbers floating around in the air are difficult for listeners to keep track of.

21. Those seedlings genetically programmed for faster growth will be selected for, whereas . . .

22. A slide with a drop of the culture was examined at 50X using a dissecting microscope.

23. Measurements of salamander respiration typically took one-half hour each.

Index